1 MONTH OF FREE READING

at

www.ForgottenBooks.com

By purchasing this book you are eligible for one month membership to ForgottenBooks.com, giving you unlimited access to our entire collection of over 1,000,000 titles via our web site and mobile apps.

To claim your free month visit:

www.forgottenbooks.com/free892765

* Offer is valid for 45 days from date of purchase. Terms and conditions apply.

ISBN 978-0-266-80923-4
PIBN 10892765

This book is a reproduction of an important historical work. Forgotten Books uses state-of-the-art technology to digitally reconstruct the work, preserving the original format whilst repairing imperfections present in the aged copy. In rare cases, an imperfection in the original, such as a blemish or missing page, may be replicated in our edition. We do, however, repair the vast majority of imperfections successfully; any imperfections that remain are intentionally left to preserve the state of such historical works.

Forgotten Books is a registered trademark of FB &c Ltd.
Copyright © 2018 FB &c Ltd.
FB &c Ltd, Dalton House, 60 Windsor Avenue, London, SW19 2RR.
Company number 08720141. Registered in England and Wales.

For support please visit www.forgottenbooks.com

STEEL TRACK

HIGHWAYS.

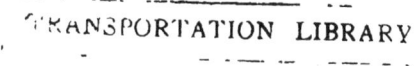

O'Donnell Steel Track Co.

New York City.

NEW YORK:
ELECRTIC PRESS
143 CHAMBERS ST.

1896

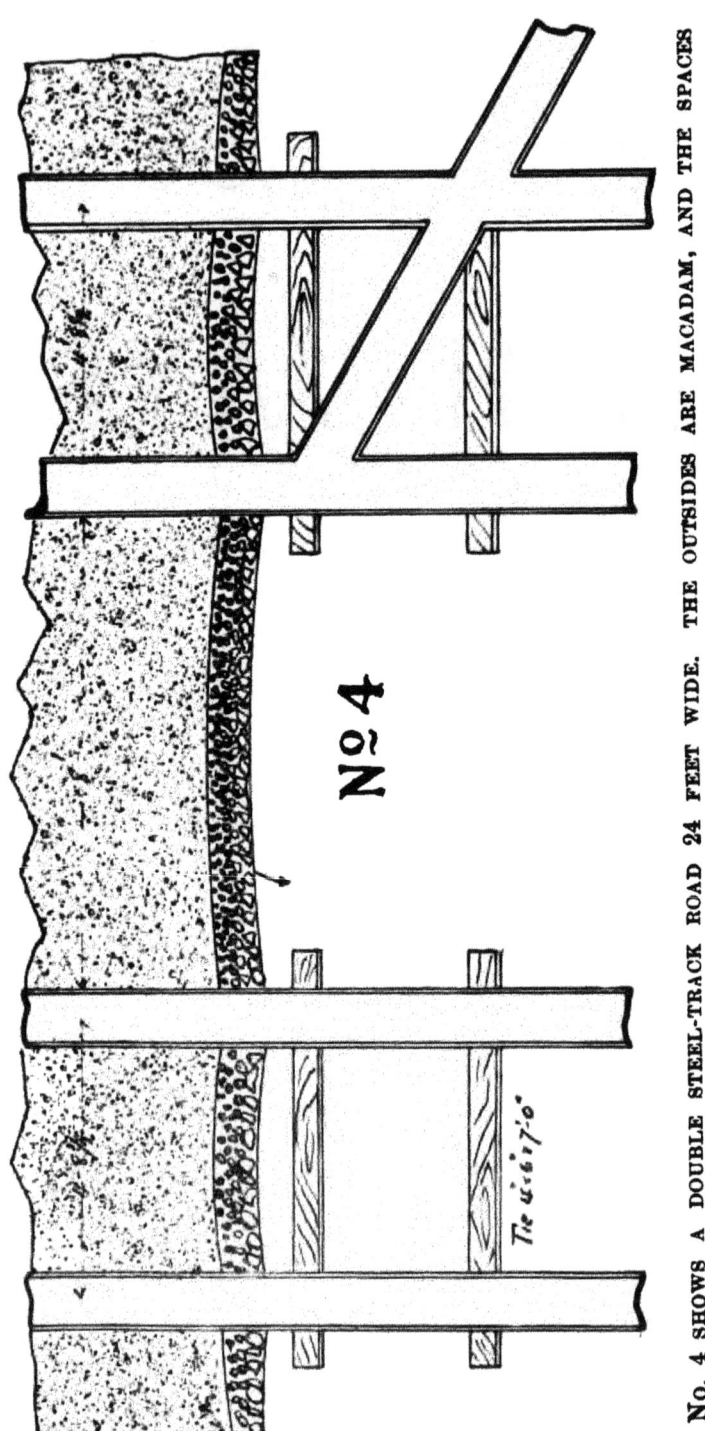

No. 4 SHOWS A DOUBLE STEEL-TRACK ROAD 24 FEET WIDE. THE OUTSIDES ARE MACADAM, AND THE SPACES BETWEEN THE 5-INCH GUTTER STEEL RAILS. THE DIAGONAL CONDUIT TAKES THE WATER FROM THE TRACK. THE CUT SHOWS THE ROAD TIES AND THE FINISHED ROAD.

THE COMING ROAD OF AMERICA.

THE COMING ROAD OF AMERICA—A STEEL TRACK HIGHWAY.

On the 26th of October, 1895, at the Road Parliament of the Atlanta Exposition, for the first time was presented to the public a model of a gutter steel track road. This Parliament, composed of road experts appointed by the Governors of the different States, in all more than one hundred and fifty, after a critical examination and extended discussion of the proposed steel track highway, passed unanimously the following resolution:

Resolved, That we recommend the construction of experimental lines of steel track highway at various points throughout the States for public travel in order that the practical value of such system may be determined.

Although not a page of literature has been published by its projector—Hon. John O'Donnell, Lowville, N.Y.—until now describing this proposed highway, he has been overwhelmed with letters of inquiry from all over the United States as to cost and where the steel track could be procured. It was impossible to answer the inquiries definitely, as negotiations were pending with steel companies on the subject. These companies very naturally hesitated to assume the large expense of making rolls to roll the proposed steel track without an exhaustive investigation of the patent (granted just about the time of the Atlanta Exposition) and the utility and public demand for steel track highways. This caused a delay of some weeks, resulting in a contract with the Pennsylvania Steel Company of Philadelphia to fill all orders for rails based on the market price of steel billets. This company have secured the exclusive right to manufacture and sell these rails in all the States east of the Alleghany Mountains, including Eastern Pennsylvania; also the right to sell in any territory outside until further notice. With this introduction we answer the question, What is a STEEL TRACK HIGHWAY?

In its general appearance at a short distance it looks like a horse railroad. Upon a closer inspection the steel tracks are gutter shape, 5 inches wide, flat on the bottom, with sides about $\frac{1}{2}$ an inch high, then extending outwards at a right angle about $\frac{1}{2}$ an inch to $1\frac{1}{2}$ inches, more or less; then downward, where it is spiked to a longitudinal timber, which makes the track perfectly smooth, with no spike heads or bolts. The track on the inside edges is slightly rounded to facilitate turning off or on the track.

HOW THE ROAD IS BUILT.

First the road bed is graded the desired width. Twelve feet is a good width for an ordinary one-track road, the steel track being placed on one side, but on a thoroughfare the road should be wider, and still wider for a double-track road, which is the model road where there is large travel. When the track is placed on one side of a road the other side is made of the same material as the filling between the track, so as to make a solid roadbed to turn out on with loaded vehicles. First the road is graded or levelled with a ditch about 8 inches deep. When finished this ditch will be about 16 inches deep. The ties are about 4 by 6 inches and 7 feet long, projecting a little beyond the track timbers, which are laid 4 feet $8\frac{1}{2}$ inches from centre to centre—the standard guage for railroad cars and all other vehicles. The ties are spaced about 3 feet apart more or less, depending upon the hardness of the roadbed. Upon these ties are laid longitudinal stringers or timbers about 6 inches wide to fill the space between the bottom flanges of the steel track, and 4 inches thick, more or less. The timbers being securely spiked to the ties, the rail is spiked to the timber 4 feet $8\frac{1}{2}$ inches from centre to centre—the standard guage for all railroads and wagons. The road is now ready to fill between the rails and at the sides with stone, macadam finish, or any other suitable material.

HOW THE SURFACE IS BUILT.

All standard authorities on road-building insist on one fundamental principle, that the rainfall must be conducted off from the surface of the road to the side drains. Macadam and Telford roads are crowning in the centre, so that the water may run off rapidly to the sides. By reason of heavy travel on these roads wagon grooves or ruts are formed where the water rests, and if not speedily filled up and repaired the road is inevitably spoiled. This constant repair of such roads by reason of wheel ruts makes their proper care very expensive, and the cost of these repairs alone will pay for steel tracks. Between the steel tracks, and at the sides, the proposed width is filled in with stone or gravel, sand or even good earth up to about three or four inches of the top, and on the top a macadam stone surface is made and properly rolled smooth and hard, about two inches crowning at the centre to the edge of the steel track, where it projects up when first made one-fourth of an inch, but in course of wear this edge will wear down level with the track. It will be readily seen that all the rainfall will run off the road into the gutter track, which in turn is conducted to the sides of the road at suitable intervals by surface side conduits about the size and shape of the rail, or by small paved stone gutters. The side of the road from the track outwards is a little depressed on the outside to conduct the water to this side of the road. The gutter track makes a perfect drainage, but without the side conduits would be worthless. The roadbed underneath will be as dry as though housed by a shed, only the natural moisture from the bottom penetrating the roadbed. This description is of a first-class steel track with macadam surface, but a

MUCH CHEAPER ROAD

can be built with steel tracks. The wood ties laid crosswise of the roads, with the longitudinal timbers on both sides, make a strong crib to hold in place any material used, which is a very important factor, for with such a crib a 4-inch surface is stronger than an 8-inch roadbed built in the ordinary way. This crib is also very important in building a road over swampy or wet places. To build an ordinary country road where there is gravel and sand, or either alone, the roadbed may be filled up to 4 inches of the top, and after being properly rolled, stone may be used to finish the surface, as in the first case, or if there is sand and blue clay the road may be filled with sand and the top surfaced with clay with a liberal sprinkling of sand, and when rolled and dry the surface will become hard, like tile, and shed rain as well as a stone road. For an ordinary cross-road the track may be only 6 feet wide, and the sides filled in with any material that will hold an ordinary wagon load. A still cheaper farm road may be built by using a flat steel track 6 inches wide. This track is spiked with a hook-head spike to the timbers with beveled wood sides about 2 inches wide spiked on to the longitudinal timbers lapping over the iron $\frac{1}{2}$ an inch on each side. Such a road can be built very cheaply if built by owners of lands or contractors along the route. The wood sides, when worn out, can be easily replaced. This cheap form of road is also used to connect a highway with a farm residence or barn by the owner, thereby practically making a steel track highway to his own door. Another important fact in building steel track roads is that as soon as the steel track is laid carts or wagons can be put on to haul material to fill. This material can be hauled by one horse twenty miles on the steel track cheaper than a team can haul the same load on a dirt road one mile. The saving in hauling material alone on the steel tracks in some localities will pay a large part of the construction.

ARE SUCH ROADS DURABLE?

The question is asked about the durability of the steel track and the track timbers. A good railroad tie on a dry surface will last from ten to fifteen years, although constantly exposed to the rain and sun, which continually disintegrates and destroys the road. The ties in a steel track road are entirely covered from the rain and sun. It is believed that the life of such ties and timbers will be more than doubled by being thus protected, and if painted with coal tar or other wood preservative their durability will be from twenty to thirty to thirty years. It is not unusual to take up a fence post or timber so protected that has been in the ground fifty or more years. Much depends upon the kind of timber, the time of year cut and whether seasoned or not when used.

The roadbed on a macadam or ordinary road is not materially injured by the tread of horses. It is the wagon wheels that do the damage. These form ruts, and although small at the beginning, hold a quantity of water, and by continual use rapidly increase in depth, and the expense of constant repairs is very large. The steel track, on the contrary, is always smooth and firm, and the roadbed being only subject to the tread of the horses, the repairs are very slight, and the entire roadbed is practically indestructible.

HOW MUCH WILL A TEAM DRAW?

Upon a level steel track a horse will draw twenty times as much as on a dirt road. This is a marvellous economic fact which hitherto has not been utilized by the public. The annual loss in hauling produce and merchandise to and from markets off the lines of railroads is simply enormous. On such tracks one horse will draw far more than two on an ordinary road. The farmer saves one horse and fixtures and his keep, fairly worth $125 per year. Then the time on the road will be reduced 50 per cent., and the "rainy spell" will not hinder going to and from market, which is virtually a saving of six weeks in the year. The advantage of being able to go to market in the "nick of time" is of immense importance, for after a "bad spell," when the roads are "settled," the market is glutted and prices go down—all of which loss falls upon the farmer.

The December Bulletin from the Road Department at Washington says:

"The national importance of this subject is fully set forth in the bulletin from the Road Department at Washington as follows: '313,349,227 tons of farm products were hauled over country roads at a cost of $663,869,000, and on this basis the cost was over 24 per cent. of its value to haul to home markets.' Commenting on these important figures, the department says: 'This increase in cost of haulage is by no means the only loss by bad roads. The loss of perishable products for want of access to markets when the market is good adds many millions to the actual tax of bad roads. The enforced idleness of millions of men and draught animals during large portions of the year can hardly be estimated. Information in the department of road inquiry indicates that nearly two-thirds of this vast loss can be saved by road improvement, and this at a total cost not exceeding the losses of four years." These cold facts ought to bring a blush of shame to the statesmen of the nation and to the several States for their almost criminal neglect of good roads.

This annual loss for a series of years, if saved, would build steel track roads over every principal highway in the United States, and add untold millions of dollars to the wealth of the people.

STEEL TRACKS COMPETITORS WITH RAILROADS.

Nor is this all. A steel track highway will be an active competitor with lateral railroads connecting with main lines, for slow freight can be hauled on a steel track road for short distances at less cost per ton per mile than on a steam road, and at less than half the average charge on lateral roads for a distance less than fifty miles. If these statements are true steel track roads mean an economic revolution. The uniform practice of all trunk or through lines, in the language of one of the highest railroad officials in the nation, is to charge "all that the traffic will bear," which literally means, to meet the fierce competition of rival routes to the seaboard the freight along lateral roads shall be charged enough more to make on the whole a satisfactory profit. This uniform practice of the trunk lines, who control all lateral feeders, has been carried to such lengths that freight to and from a distance of a thousand miles is carried to and from market at less per ton than is charged on lateral roads for a hundred miles. The following instances out of a great number will illustrate the great wrong done to the producers and consumers of the United States: Gen. Curtis, now Member of Congress from St. Lawrence, N. Y., declared on the floor of the Assembly, of which he was then a member, that he could ship a barrel of onions from his farm to Chicago and from there to New York cheaper than he could ship direct by rail to that city, although the distance *via* Chicago was more than fifteen hundred miles more. In an examination held in New York City in September, 1995, to reduce the rates on milk, on certain railroads leading into the city the railroad officials testified that they charged as much per can for milk shipped into the city three miles as for three hundred. There is not a State in the Union where similar instances of oppression do not continually occur.

SLAVERY OF SHIPPERS TO RAILROADS.

The result of such a policy is to enslave the farmers and manufacturers of the country. The most absolute tyranny in the fixing of rates has been the rule of railroads, particularly on lateral roads. On main lines pools and trusts usually fix prices. These are broken by quarrels or otherwise when it is to the interest of the conspirators. Under this oppressive policy of the railroads the natural law of geographical "nearness to market and increased value of farm lands" has been reversed, and as a consequence farm lands in the State of New York have declined in value at least 50 per cent. in price since the constructing of trunk lines through the State. This decline began upon their completion, but was arrested in part during the war; but since then the decline has gone on with accelerated pace, until the farmers of this State are practically bankrupt. And the same applies to all other States where trunk railroad lines control, particularly in the South and West.

THE MASTER AND THE SLAVE.

Before the War the slave toiled for his master, and his imperious will was law. To-day the farmers and business classes are in abject slavery to the railroads. In all other business the buyer and seller are on an equality. If either are dissatisfied with the terms of the other there is no compulsion on the part of either. How is it with the shipper by rail? Let him remonstrate against rates or try to reason with his master! He is promptly told "these are our rates,' and although they may be ruinous to the shipper he has no alternative. Was ever slavery more abject? Washington said, in his farewell address to his countrymen, "Upon the prosperity of agriculture depends the prosperity of the nation." This great economic truth was re-echoed by Jefferson, Madison and other early statesmen. The country is not suffering to-day from the currency or tariff, but the one all-important cause is the depression of agriculture through enforced slavery to railroads.

With these damning facts before statesmen, State regulation of railroads in the interests of shippers was devised. Georgia led some years ago with a State Commission, and every State followed. But State commissions were powerless to redress the wrongs suffered by the people. Then the strong arm of the nation was outstretched in a National Commission representing the power of the nation. But the majority of the first Interstate Commission was composed of former railroad attorneys. The patent fact is that these powerful corporations not only own and control State legislators, but the nation's also.

STEEL TRACK ROADS THE REMEDY.

At this juncture in the history of the nation the only effectual and secure remedy is the building of steel tracks on the common highways free to every man, without toll or tribute to pay interest on grossly watered railroad stocks and bonds. The immense savings to the people by steel track roads is illustrated by the following: At a point, on a connecting lateral road with the New York Central twenty teams in the fall of the year delivered twenty loads of cheese from a factory twelve miles distant from the railroad. The average load was 1,600 pounds. It took all day to deliver the 32,000 pounds of cheese and return. On a steel track road that cheese could have been delivered by one team in less time and in better condition, thereby saving nineteen teams and nineteen men, fairly worth $57. But this is not all. This cheese was shipped to the connecting point on the Central, fifty-nine miles, at twenty cents per hundred—$64. On a steel track road three teams would have delivered it and returned in three days at $3.50 per day, or $31.50, saving $32.50, and if loaded back at the same rate the total saving would be $65.00. The market price of the cheese at eight cents per pound was $256, and the loss by reason of roads and rail charges was over one-third the price obtained. In any State in the Union this statement can be matched by actual facts. To this immense loss is to be added the increased freight on every pound consumed in the family, farm or in the avocations of trade and commerce. In the light of these illustrations, is it any wonder where the profits of the farm goes, and that farm lands decline in value?

The question is asked, will not steel track highways injure railroads? One fact is of more importance in an argument than a hundred theories. When the elevated railroads were first projected over the surface street roads in New York City they were opposed by the surface road stockholders on the ground that the elevated roads would carry all the passengers, and for a short time these stocks were depressed But the elevated roads were built over every principal avenue in the city on which were surface roads, and for years they have carried over 600,000 passengers daily, and surface railroad stocks were never before at so high a price in the market, although the fares have been reduced fifty per cent, and with all these facilities New Yorkers are clamoring for an underground road to relieve the travel on the surface and elevated roads. The universal law is that "increased facilities create increased travel," and steel track roads will not be an exception. On the contrary, while comparatively slow freights and travel will take the steel road, business all along railroad lines will be stimulated and increased, and these steel track roads will be constant feeders to the railroads. So apparent is this that the most astute railroad managers have encouraged the building of good roads wherever projected.

The steel track road, beside its general use for wagons, is the best road yet devised for bicycles. The 5-inch steel track on either side will do for racers, while the smooth macadam road between the tracks will make the finest tracks possible. The coming horseless motors will also find this the very best road that could be possibly made for their use. This track is also the very best electric motor track, and serves admirably for a motor that runs but once or twice a day to and from given points. Such motors are now being constructed that will turn on and off such a track as easy as a wagon. This will also make an admirable light railroad track. In many country towns a light locomotive could be owned by the people and run over the road back and forth once or twice a day and haul one or two cars at an astonishing small cost—very much less than an electric motor. It is believed on a steel track road owned by the people, in addition to the ordinary wagon use, passengers can be carried twenty miles an hour at less than a quarter a cent per mile.

STEEL TRACK ROADS A GOD-SEND TO FARMERS.

The uniform experience of farmers and real estate owners along the line of improved stone roads in the States of New Jersey and Massachusetts proves that the price of land has wonderfully increased since their construction. Here, again, one fact outweighs a hundred theories. In fact, as one writer expresses it, "the towns are literally tumbling over each other to see which will get there first."

If such marked increase in values have taken place since the construction of these stone roads, what will be the increase when steel track roads are built? There is little doubt that on these stone roads the steel track will be ultimately laid, for the saving in expense of repairs in a few years would pay for the steel rails.

Macadam and Telford roads date back a hundred years to the stone age—the Appian way and the ponderous Pyramids. But this is a wonderful steel age. Almost as by magic we have steel suspension bridges, tunnels, towers, aqueducts, lift and canal locks, sky-scraping buildings, and, last and most important of all, steel track highways for the common people. One uniform system of steel tracks is to span the Union, binding the North and South and East and West together in friendly bands of steel. No matter whether the road be a thoroughfare costing thousands of dollars or the humble farm road costing only hundreds, the steel track

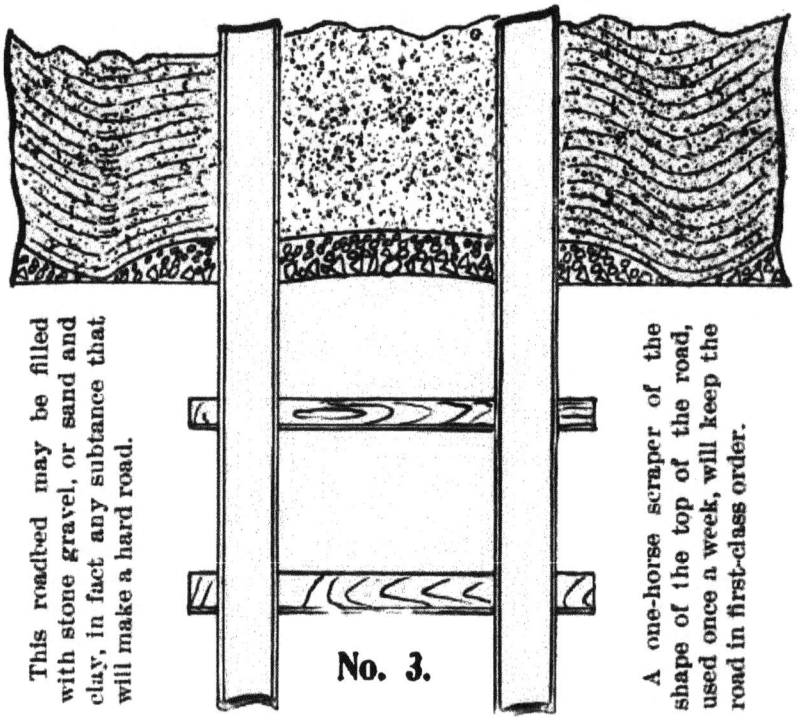

NO. 3 SHOWS A CHEAPER FARM ROAD. THIS MAY BE 6 OR MORE FEET WIDE, WITH THE GRADE ON ONE SIDE AS IN CUT 2, OR WITH GRADE ON BOTH SIDES, AS HERE SHOWN.

in shape and capacity will be the same, so that the citizen with his own horse and vehicle or motor can ride from one end of the continent to the other without "change of cars or baggage," on a road as free as the air, owned by the common people and without toll or tribute to any lord or master.

To the West and South, in particular, steel track roads will be the harbinger of better days. These roads, on the level prairies, can be built not only very cheap, but in an incredibly short time. On these roads grain can be delivered in any weather and at all seasons of the year, thus preventing a glut in the market and the combinations of the rail and elevator whereby millions of dollars are annually lost. Wheat and corn can be delivered at less than half the freight now charged, and the farmer will not only save his cash freight in the employment of his own team, but long wagon train loads of grain can be delivered by mutual arrangement of farmers along the line, thus combining the producing interests in defense against rail monopolies, who take all the profits of the farm, leaving the slave-farmer only barely enough for his subsistence.

Steel track roads mean a boom in farm lands, for they invite back the flower of our sons and daughters who have left the old farm for the social advantages of the village or city, and the reflex influences will alike elevate city and country, and make the country the favorite residence place for the best and greatest of the city and nation.

Steel track rural roads will not only make the farmer independent, but, as heretofore stated, boom his lands and bring them into market. There can be no question that the "hard times" we are passing through are directly traceable to the "handicapping of the markets," so forcibly stated by the Chamber of Commerce, and no matter what the action of political parties may be on the silver or tariff questions, no better times will be experienced by the farmers of America until the country is delivered from their bondage to the railroads by the steel track highway—the coming road of America.

WHAT FARMERS SAY.

We have selected the report of the Commissioner of New Jersey for illustration, not having space for the others. We give below not the Commissioner's own report, but the report of the farmers of New Jersey in answer to a circular sent out by him asking their opinions of the State Aid law and of good roads. Here is a sample of the answers:

(From H. Darnell, Mount Laurel, N. J.)

I would say that it is the universal opinion of farmers that our agriculture has been benefited more by our State aid in building good roads than anything that has ever been done for them before.

(From Samuel J. Allen, Cinnaminson, N. J.)

At first the farmers of the neighborhood were largely opposed to the movement, but within the past year they have become almost unanimous in favor of the extension of the system. I am hoping that the State appropriation may be largely increased.

(From H. H. Brown, Old Bridge, N. J.)

We have four miles of Macadam road through our township. Property has almost doubled in value, and travel has more than doubled. Farmers carting over our road speak in the highest terms of it. It is a rest for their horses to cart over it. Our citizens are greatly in favor of having it extended.

(From Charles Vauscuer, Beverly, N. J.)

In reply to your request for my opinion, I am glad to say I

consider the law most beneficial to farmers, as it is the farmer that uses the roads, and his labors are more or less burdensome in proportion to the condition of the roads.

(From Clayton Conrow, Cinnaminson, N. J.)

Goods roads enable the farmer to place his products on the market when at the highest point; bad roads often forbid this. Good roads enable him to deliver perishable vegetables and tender fruit in so much better condition that they invite the better class of customers and command a higher price. We can't afford bad roads. Their effect is enervating, while that of good roads is inspiring.

(From Dennis Long, Union, N. J.)

I think the law granting State aid is one of the best that can be put on our statute books. We have a county road in our neighborhood of which we are very proud, and near which property advanced 50 per cent. of its value while the road was laid. A case in point is that of a piece of property along this county road which could have been bought for $20,000 before the road was built, and only a few days ago the owner was offered $30,000 for the farm, and it is two miles away from any railroad station. And there are many more cases of the same sort.

(From Samuel C. DeCou, West Moorestown, N. J.)

From about two years' observation and experience on a good stone road, I think they are very hard to overestimate, either from a business standpoint or pleasure standpoint. Teams cannot only draw double the loads formerly drawn on ordinary gravel roads, but on level or descending grade they actually rest. Besides the economy of increased loads, we are no longer compelled to make deductions for very wet weather, very dry weather or spring time, when the frost is coming out of the ground. The State has very wisely lent a helping hand in this important matter.

(From Howard G. Taylor, Riverton, N. J.)

The farmers in this section are very much in favor of State aid for improved roads. A good stone road enables us to haul larger loads at less expense for teams.

(From Joseph A. Burroughs, Merchantville, N. J.)

Farmers are greatly benefited in horseflesh, time and expense by the aid of the State in building stone roads. We can do with two horses what it required four to do.

(From Manning Freeman, Merchantville, N. J.)

I have only to say that farmers are very enthusiastic for good roads. I would not sell my house and accept another worth $7,000 as a gift and be obliged to live in it if two miles from a macadam road. No farmer in this neighborhood would buy a farm not located on the macadam road. Now that they have a sample of the road, they all want it.

STARTLING FIGURES.

The possibilities of steel track roads on the level prairies of the West and the deltas of the South will be seen by the following: One horse will draw on a steel track twenty tons, or over 700 bushels of corn. But to make a conservative estimate call it 500 bushels. A horse will draw this load and return thirty miles in two days. The wages of horse and driver call five dollars, or one cent a bushel, for the thirty-mile haul. A larger load and a longer distance can be drawn as cheap or cheaper per bushel. Take a distance of 100 miles. Place three horses abreast—one between the tracks and the other two outside the tracks. On the same basis three horses will draw 1,500 bushels, or 42 tons, but to make a very conservative estimate call it 1,000. The three horses will travel the hundred miles in six days at a cost of $10, making the cost of moving 10,000 bushels of grain 100 miles $10, or one cent per bushel. It may be asked how would you load the grain. The answer is, just as it is now, loaded on wagons. These would form a train like a train of cars, and the train might start from the most distant point, taking on wagons at every farm, stopping just as cars stop now at way stations. These trains will draw a trainload about as easy as though the load was on one car, the friction being but a trifle more. In these calculations nothing has been said about a return load, which in the shipment of coal, groceries and general merchandise will far more than pay the entire cost of haulage one way, and at a very large per cent. lower than it now costs the consumer either by rail or teams.

These calculations have been made for delivering grain on long hauls, but the difference in favor of steel track roads for the average haul to reach the railroads that now do the shipping is still more marked. Agricultural Bulletin No. 19, from the department at Washington, November, 1895, has the following interesting report:

"With the aid of the Division of Statistics of the Department reports have been gathered from about 1,200 counties giving the average length in miles, from farms to market or shipping points, the average weight of load hauled and the average cost per ton per mile, and from this is deduced the average cost per ton for the whole length of haul.

"These returns have been arranged in groups of States, and the result shows that the average length of haul in the Eastern States is 5.9 miles; in the Northern States, 6.9 miles; in the Middle States 6.8 miles; in the Cotton States, 12.6 miles; in the Prairie States, 8.8 miles; in the Pacific Coast and Mountain States, 23.3 miles, and in the United States 12.1 miles.

"The average weight of load for two horses in the Eastern States is 2,216 pounds; Northern States, 2,136 pounds; Middle-Southern States, 1,869 pounds; Cotton States, 1,397 pounds; Prairie States, 2,409 pounds; Pacific Coast and Mountain States, 2,197 pounds, and the United States 2,002 pounds.

"The average cost per ton of 2,000 pounds per mile in the Eastern States is 32 cents; Northern States, 27 cents; Middle-Southern States, 31 cents; Cotton States, 26 cents; Prairie States, 22 cents; Pacific Coast and Mountain States, 22 cents, and the United States 25 cents.

"The average total cost per ton for the whole length of haul: Eastern States, $1.89; Northern States, $1.86; Middle-Southern

States, $2.72; Cotton States, $3 05; Prairie States, $1 94; Pacific Coast and Mountain States, $5 12, and the United States $3.02.

"With these data it becomes possible to obtain approximately the total cost of the entire movement of farm products and other classes of materials over country roads.

"The total farm product gives 313,349,227 tons, which, at the average cost above stated, $3.02 per ton, makes a grand total for the annual cost of haulage on the public roads of $946,414,662.54.

"The immensity of this charge will be best realized by comparing it with the values of all farm products in the United States for the year 1890—$2,480,170,454—which values have probably diminished since that date.

"The increase in cost of haulage actually done is by no means the only loss by bad roads. The loss of perishable products for want of access to market; the failure to reach market when prices are good, and the failure to cultivate products which would be marketable if markets were more accessible, add many millions to the actual tax of bad roads. Moreover, the enforced idleness of millions of men and draught animals during large portions of the year is a loss not always taken into account in estimating the cost of work actually done.

"Information already in possession of the Office of Road Inquiry indicates that, all things being considered, nearly, if not quite, two-thirds of this vast expense may be saved by road improvement, and this at a cost not exceeding the losses of three or, at the most, four years by bad roads."

Commenting on these startling figures, the Chamber of Commerce of the State of New York, the oldest and most important commercial body in the United States, says:

"The movement for good roads deeply concerns every commercial and financial interest in the land. We are handicapped in all the markets of the world by an enormous waste of labor in the primary transportation of our products and manufactures while our home markets are restricted by difficulties in rural distribution which not infrequently clog all the channels of transportation, trade and finance."

COTTON IS KING.

In the years gone by cotton was king, and wielded his imperial sceptre all over the world. But cotton to-day is a dethroned monarch, and although one of the world's great factors, yet his primal glory has departed. In the years gone by the fertile fields of the South supplied the spindles of the world, and the transportation of the products of the plantation was a tempting prize to the barons of the rail. Railroads sprung up as by magic all through the South. These roads would have been real benefactors to the people had they been run on honest, sound, business principles. But when built they were capitalized at three or four times their cost—literally watered to death. And a mad strife to make the staples of the South—cotton, sugar, rice and tobacco—pay a large interest on this fraudulent capital began, and has continued to this day. What has been the result? The enhanced price of cotton in the market on account of the cost of transportation stimulated competition in growing these staples in other countries until cotton could be placed in the marts of the world at prices ruinous to the planters of the South. To-day the South has partially recovered her prestige in the numerous mills built to manufacture the cotton, but the one great and almost unsurmountable obstacle to the prosperity of the South is the exorbitant freight rates of the railroads.

Here are a few examples which can be duplicated wherever cotton is shipped. At a point in the State of Mississippi there is a railroad to the river, eight miles long. On this road the lowest the planter can deliver his cotton is 15 cents per hundred, and the regular rate for a distance of from 50 to 75 miles is from 25 to 40 cents per hundred. In South Carolina the regular rate for shipping cotton

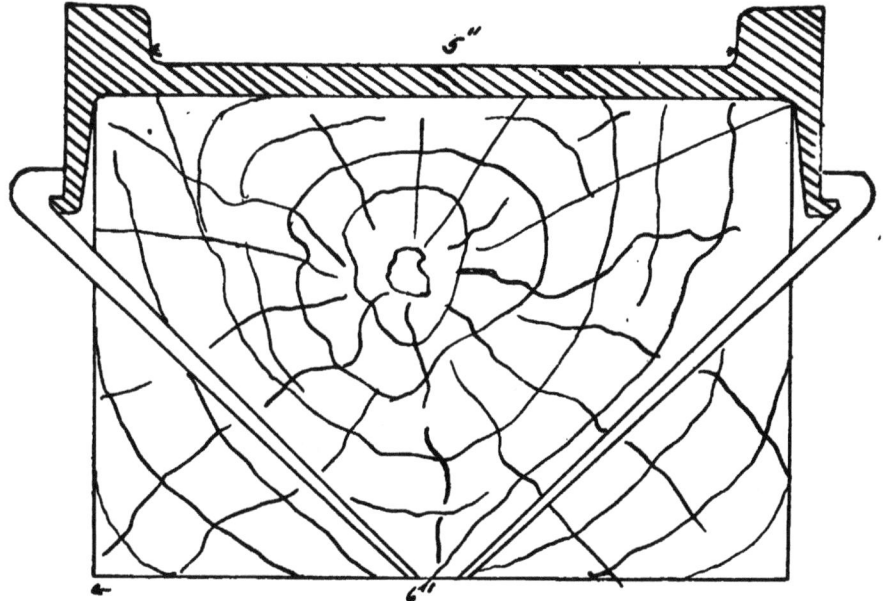

NO. 1 SHOWS AN END VIEW OF THE TRACK AND LONGITUDINAL TIMBER WHICH RESTS ON THE CROSS TIES. SMALL WOOD WEDGES FILL THE SPACE NEXT TO THE TIMBER BACK OF THE SPIKE HEADS.

is from 40 to 50 cents per hundred for a distance from 50 to 100 miles.

The fruit growers of the South are literally robbed by transportation companies; only enough is left to the grower to keep soul and body together. The planter who paid 15 cents per pound to carry his cotton 8 miles, on a steel track highway could have carried it at less than one quarter the charge, and cotton can be moved anywhere on and off the plantations at much less than half the present cost.

Steel track roads will make the planter master again, and the lord of the rail will have to surrender his sceptre to the rightful owner. Steel track roads will add millions on millions to the wealth of the

South as well as to the West. The motive power may be the horse or the mule or a light locomotive, as the road can be used by either, thus uniting both the primitive and the modern economic forces for the service of man.

THE COAL BARONS.

The coal barons of the land are in league with the railroads and the railroads fix the price and output in as tyrannical a manner as the barons of old tolled their subjects. The coal fields, both in the mines and in transit to market, will receive an immense benefit from the steel track railroads. While the working of the mines will be greatly facilitated by steel tracks for hauling, it remains to be seen if the steel track is also to be the great deliverer of the people from the pools, trusts and combines that now put up the price at their own sweet will to the consumer while they, at the same time, put down wages to the men who delve in the mines. Coal can be carried to many points of consumption at half the prices now charged, and with the completion from county to county of steel track roads the glad era of cheap coal will begin, and as a consequence all the activities of manufactures and commerce will receive new power and importance.

UNCLE SAM'S WEALTH.

The vast unoccupied domain of the Government, consisting of more than 600,000,000 of acres, waits the magic development of steel rails. Nothing could inure so much to the benefit of the Government in the rapid settlement of this vast territory as the building of steel track highways to open up this last heritage of the people saved from the voracious maw of the railroads. These lands are distributed in the following States and Territories: Arizona, California, Colorado, Idaho, Montana, Nebraska, Nevada, New Mexico, North Dakota, Oklahoma, Oregon, South Dakota, Utah, Washington, Wyoming. The general government has practically donated hundreds of millions of dollars to the Pacific railroads. And these roads have turned upon the people and in high freight and passenger rates crippled their industries. And they have become so powerful that they not only repudiate their honest obligations but defy the sovereign power of the people. Here is a problem for the statesmen of the West now in Congress. Why should not the general government aid in the construction of steel track highways in these States? Every argument that applies to State aid in building roads applies with tenfold force to Government aid in development of Government lands. That the investment would be immensely profitable to the Government in the peopling of the States needs no argument, and the individual States could very profitably join in the general movement to build steel track highways. Where are the statesmen of the West to champion such governmental aid for their respective States?

THE AMERICAN ITALY.

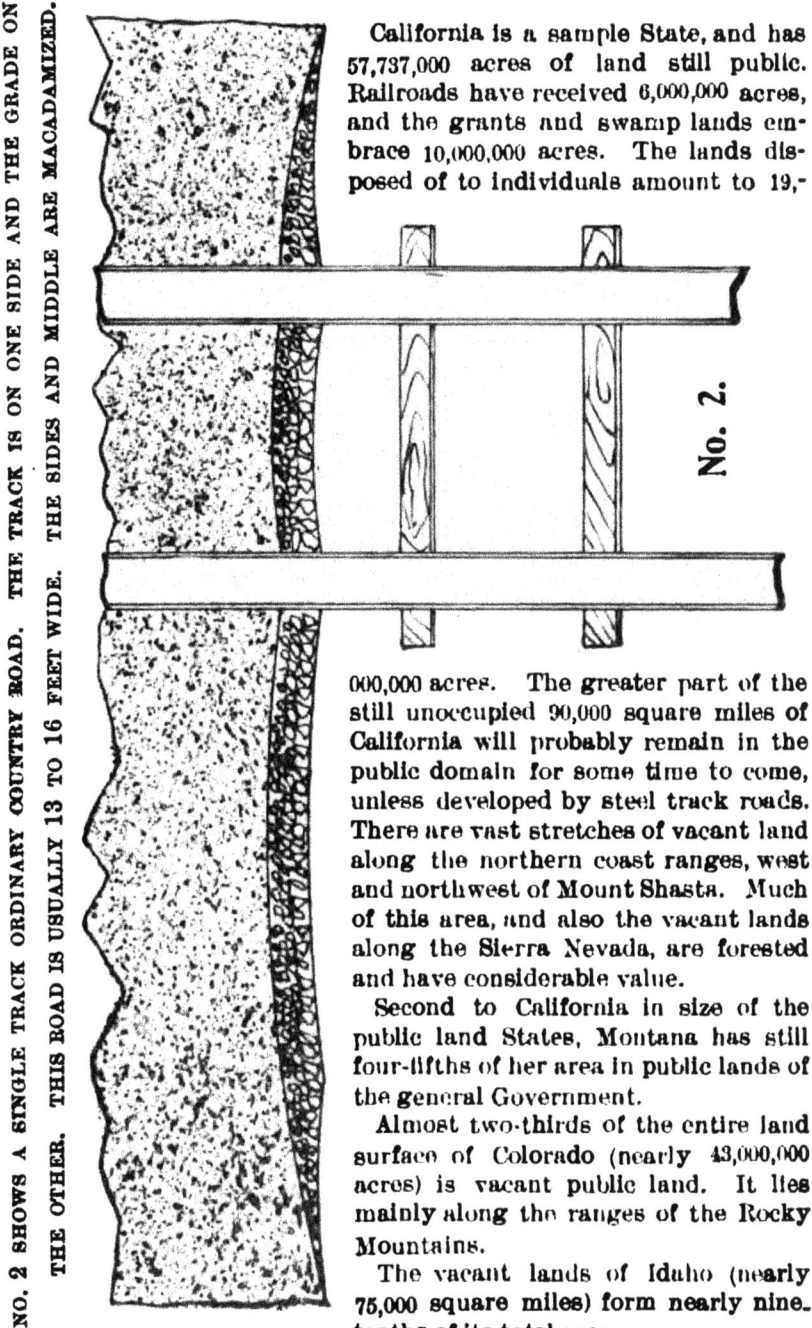

NO. 2 SHOWS A SINGLE TRACK ORDINARY COUNTRY ROAD. THE TRACK IS ON ONE SIDE AND THE GRADE ON THE OTHER. THIS ROAD IS USUALLY 13 TO 16 FEET WIDE. THE SIDES AND MIDDLE ARE MACADAMIZED.

California is a sample State, and has 57,737,000 acres of land still public. Railroads have received 6,000,000 acres, and the grants and swamp lands embrace 10,000,000 acres. The lands disposed of to individuals amount to 19,-000,000 acres. The greater part of the still unoccupied 90,000 square miles of California will probably remain in the public domain for some time to come, unless developed by steel track roads. There are vast stretches of vacant land along the northern coast ranges, west and northwest of Mount Shasta. Much of this area, and also the vacant lands along the Sierra Nevada, are forested and have considerable value.

Second to California in size of the public land States, Montana has still four-fifths of her area in public lands of the general Government.

Almost two-thirds of the entire land surface of Colorado (nearly 43,000,000 acres) is vacant public land. It lies mainly along the ranges of the Rocky Mountains.

The vacant lands of Idaho (nearly 75,000 square miles) form nearly nine-tenths of its total area.

All this immense territory is an open field ready for the magic influence of the steel track highway to bring it in touch with civilization, trade and commerce.

STEEL ROADS FOR FORESTS.

Hon. B. E. Fernow, chief of the United States Bureau of Forestry, Washington, in a recent article, called attention to the immense benefit that would be derived from good roads in connection with forestry. Wood chopping is the primitive mode of treating the virgin forests when the pioneer enters on the woodlands. The logger robs the virgin forests of the best logs and abandons the rest. He cuts for the market and has to use the streams as runs in high water to get his logs to market. Only the logs of light coniferous and other soft woods can be so transported. Hence, much of the most valuable timber of the hardwood varieties must remain standing. With steel track roads in the forest all kinds of timber could be regularly delivered in market when wanted. The chief says: "The little city of Goslau in the Hartz mountains of Germany owns a forest of 7,500 acres. Until 1875 the district was without proper roads, but in that year the city appropriated funds to build first-class roads. In 1891 there had been spent $25,000 and 141 miles of road had been improved. The forest manager had kept a close account of the net result, and in that year reported 33 per cent. gain on the amount of the investment. On one particular road, which was macadamized and maintained for one year at the cost of $7,440, the following comparison of the cost of hauling the 470,000 cubic feet of wood that had to pass over it was made: On the old road this required 4 273 loads of 110 cubic feet average at $3 60, equal to $15,282.80; on the improved road the whole quantity could be moved in 2,652 loads of 177 cubic feet average, figured at the same price, or altogether $9,547.20, resulting in a total saving in haulage alone of $5,735.60, or 77 per cent. of the cost of the road in one year." Steel track roads will do for the lumberman all the year round what the spring freshet does for them in the running of logs in the spring. Besides this, millions of feet of most valuable hard wood, now left on the land, could be delivered to market regularly at prices which would more than compensate for the outlay in building the steel track roads. If the land, when the timber is removed, is fit for agricultural purposes, the steel track road will bring it at once into market, and at prices which will pay the cost of the road many times over.

STEEL TRACKS FOR CITIES.

The argument presented in favor of steel tracks for country roads in relation to traction, and the consequent power of horses to move freights easily and cheaply by ordinary vehicles, applies with tenfold force in cities. Traffic in New York City, where this article is written, in the business parts of the city, is congested and obstructed in the transportation of merchandise from one part of the city to another, blockades being frequent and the loss of time great. The loss of time by reason of the congested condition of the streets amounts to a large sum daily, while the general damage to business, vehicles and horses by reason of the condition of the streets is beyond measure. On many of the business streets railroad corporations occupy the center of the streets, and carts and wagons seek these tracks for the easier traction they present, notwithstanding they are ill-suited for the purpose, and in some cases designed to make it difficult and dangerous to thus use them. If steel tracks designed for the use of ordinary vehicles were laid, it would tend to keep teams off the railroad tracks, facilitate both passenger and merchandise traffic, double the life of existing pavements, reduce the noise of stone pavements to a minimum, reduce the number of horses, for one horse will pull much more on a steel track than two can on an ordinary pavement, equalize traffic by keeping the movement of merchandise out of passenger streets, and realize the maximum of economy, durability and comfort. Belgian or other pavements are frequently broken and rutted by the inevitable wear of the heavily loaded wagons which continually thunder over these pavements.

One horse would, on steel tracks, draw more than five on stone pavements. This would take off probably one-third of the horses, thereby saving a very large sum; reduce the congestion and facilitate the movement of freight, so that on such a street as Chambers, and other streets, there would be no blockades as at present. On a street like this street there could be only a right and left-hand track, but on streets with no railroad, and on the avenues and water-fronts, there

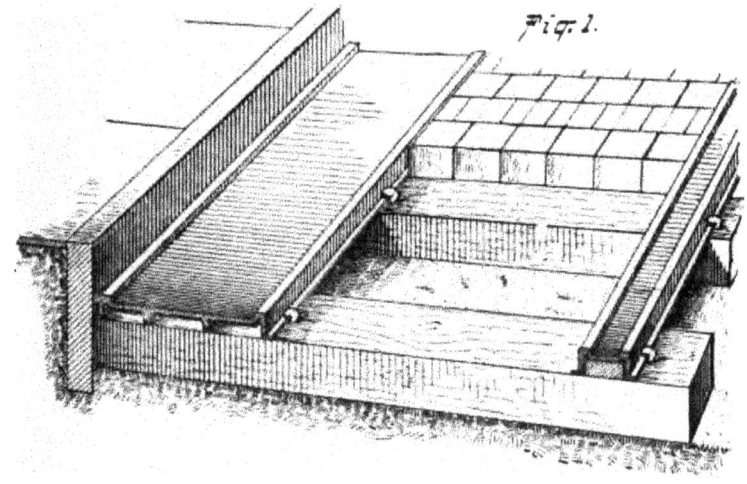

Fig. 1.

could be three or four tracks. On Broadway it would not be practicable to lay but one right and one left-hand track. On Fifth avenue a model road would be a steel track on either side (leaving room for carriages at the curb), on which heavy vehicles like furniture and express wagons who must use this street more or less, could travel, the carriage center being filled in with asphalt or wooden blocks. Such a road would be noiseless for trade and a little paradise for drivers of carriages. On all streets where steel tracks are laid there would be but little noise, which is due to the pounding of the wheels over the stone pavements. On new streets and in the suburbs the first cost of steel track streets would be no more, but actually less, than the usual contract price for pavements, taking into account their durability.

HOW STEEL TRACKS SHOULD BE LAID.

All propositions to lay steel tracks in cities have been met with the fact that there is no standard gauge for wagons or carts. While all carriages and many wagons and carts are built the standard gauge of 4 feet 8½ inches, the heavy wagons and carts are generally from that gauge up to 6 feet. This difficulty has been surmounted by making one track wider than the other – the narrow track being 5 inches wide and the other wide enough to accommodate the 6 feet wagons. The wide track has small, ⅛-inch depressions a few inches apart so that horses will not slip if they step on the track. The track is a gutter track to carry off the water, having raised sides half an inch rounded on the edge so as to make no obstruction in turning off or on. These tracks are made of steel plate from ¼ to ½ an inch thick, laid on stringers placed on cross ties (or the stringers may be omitted), the middle between the tracks and also the sides filled in with stone pavement, asphalt or wood. A steel track of this kind would in the end be much cheaper than any other kind of pave-

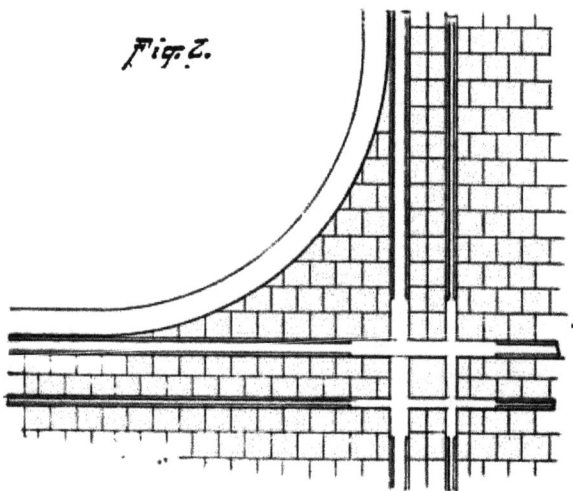

Fig. 2.

ment because of durability. One has but to look at the best paved streets of the city to find ruts and holes in all stages of decay made by the wheels of vehicles. Once a small hole in a pavement is made, the successive blows of heavy wheels, catapult like, soon wear out the best pavement. The wear of the pavement on the outside of car rails, where vehicles have used the car track, is obvious all over the city. The fact that heavily loaded teams seek to get one wheel on the car rail is a mute appeal to the city to provide two steel tracks for the use of its trade and commerce. At street crossings the flange is left off the gutter tracks. The tracks at these crossings are level and flat, with the surface inclining up half an inch from the gutter track.

The approaches from the ferryboats should have steel tracks. It is not an uncommon sight, particularly at low tide, to see a tackle used to haul loaded drays from the ferryboats. The Brooklyn Bridge and its approaches need the relief of steel tracks. The vibration caused by a loaded wagon striking the edges of the plank is much greater than that caused by the passage of the cars. An eminent engineer said to the writer that while standing on a tower of the Suspension Bridge at Niagara Falls he noticed that there was more vibration when a loaded wagon drove over than when a train of cars passed. The vibration is more destructive to the suspending cables than any other cause.

The idea of steel tracks for vehicles is not new. When very heavy loads are drawn, at iron foundries and rolling mills, steel plates are frequently spiked on planks or timbers when a very heavy haul is to be made up a steep grade. In Johnstone, Scotland, there is a steep hill where iron tracks have been laid for twenty years. Many streets along the docks in Liverpool, England, where there is heavy haulage, have long slabs of granite laid end to end in four parallel rows for the wheels of vehicles, the spaces between being paved so as to make a strong footing for the horses. With steel tracks, as proposed, laid on the river fronts of New York and from the Battery to Harlem, millions of dollars would be saved to the commerce of the city.

The granite blocks of Fifth avenue are an intolerable nuisance, even when first laid, but by continual use, as is the case with all stone pavements, they have become rutted and worn, until wagons and carriages pound out one rut into another with no relief except as a turn is made off the street on to a cross street asphalt pavement. Fifth avenue is the one great promenade of the city—the street sought out by strangers visiting the city as well as by the people of New York City. Boston has its splendid Commonwealth avenue, restricted to light driving, although in the heart of the city. Chicago has forty-five miles of parkway, mostly of asphalt, restricted to pleasure and light driving. New York has no such avenue for pleasure driving and quiet enjoyment. So far as the houses are concerned, Fifth avenue has the finest mansions of any street in the world, and yet the streets are worse than any other principal pleasure street in the world. Steel tracks laid as proposed, with asphalt, would at once make it a magnificent street of which the city might well be proud. The steel track is the complement of asphalt, and the one system wanting in an asphalt road. Asphalt, like macadam roads, inevitably rut. On a rainy day the surface of the best asphalt roads will be seen to be full of depressions caused by the wheels of vehicles. These are filled with little pools of water, and on such days the wear by driving is greatly increased. A right and left hand steel track on Fifth avenue would provide for all necessary travel of heavy wagons, which must be used more or less in the delivery of coal, furniture, express, etc., and these might be limited to the forepart of the day, leaving the avenue free the rest of the day. The stages would, of course, take the steel track, and also the major part of the carriages, because of the ease and pleasure of riding on a track as smooth as glass and with no obstructions by reason of spike or snake heads. Thus the life of the pavement would be trebled, and the cost in the saving of repairs would in the end be far more than to pay the increased cost of rails.

The annual charge upon New York for pavements is $2,000,000 a year. Granite pavement costs about $3.75 a square yard on a concrete foundation. An enlarged view of the best Belgian pavement shows a succession of little hills. Every time wheels pass over such a road the descent from one hill or block to the next is like a heavy blow from a hammer or catapult, which enlarges the depression and makes the hill larger. There is no such thing as "wearing the road smooth," for the road is best when first finished, and the process of decay begins when the first vehicle drives on the street. Was ever a more unscientific road devised? Steel tracks would make a perfetly smooth trackage which would outlast at least four pavements. It is not the horses' feet that do the main damage. This wear is but a small item compared with the wear of wheels. The saving to the city by the use of steel tracks during the life of the pavements now in use, if generally adopted, would save millions of dollars in taxes and millions more in repairs on trucks, wagons and carriages, and in the new facilities for the trade and commerce of the city. (See page 23.)

STEAM MOTORS FOR STEEL TRACKS.

The possibilities of cheap steam motor service on steel track highways in the carrying of passengers and freight at a very low rate are immense. The capitalization of the trolley roads in round numbers is over $100,000 per mile. On this vast capitalization the

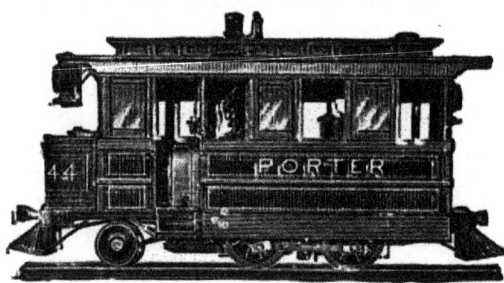

WEIGHT, 9,000 LBS.; HAULING CAPACITY, 175 TONS.

FUEL, 1 TON OF COAL FOR 100 MILES.

COST $1,200

public are charged rates high enough to pay large quarterly dividends to the stockholders. The people will be the stockholders in a steel track road, and may run trolley cars on their own road at one-half the present rates. The steel track road is also well adapted for a light steam engine. The cost of running such a steam motor capable of hauling fifty to one hundred tons of freight a hundred miles may be stated in round numbers not to exceed $10 a trip, as follows: 2,000 pounds coal, $5; engineer, $2.50 per trip; other expenses (oil, repairs, interest, etc.), $1.50. This would pay all the expense of a light locomotive, weighing 10,000 tons, haauling two to four cars on a train. The cost will vary according to price and quality of fuel, rates of wages, grades and other conditions of service. As com-

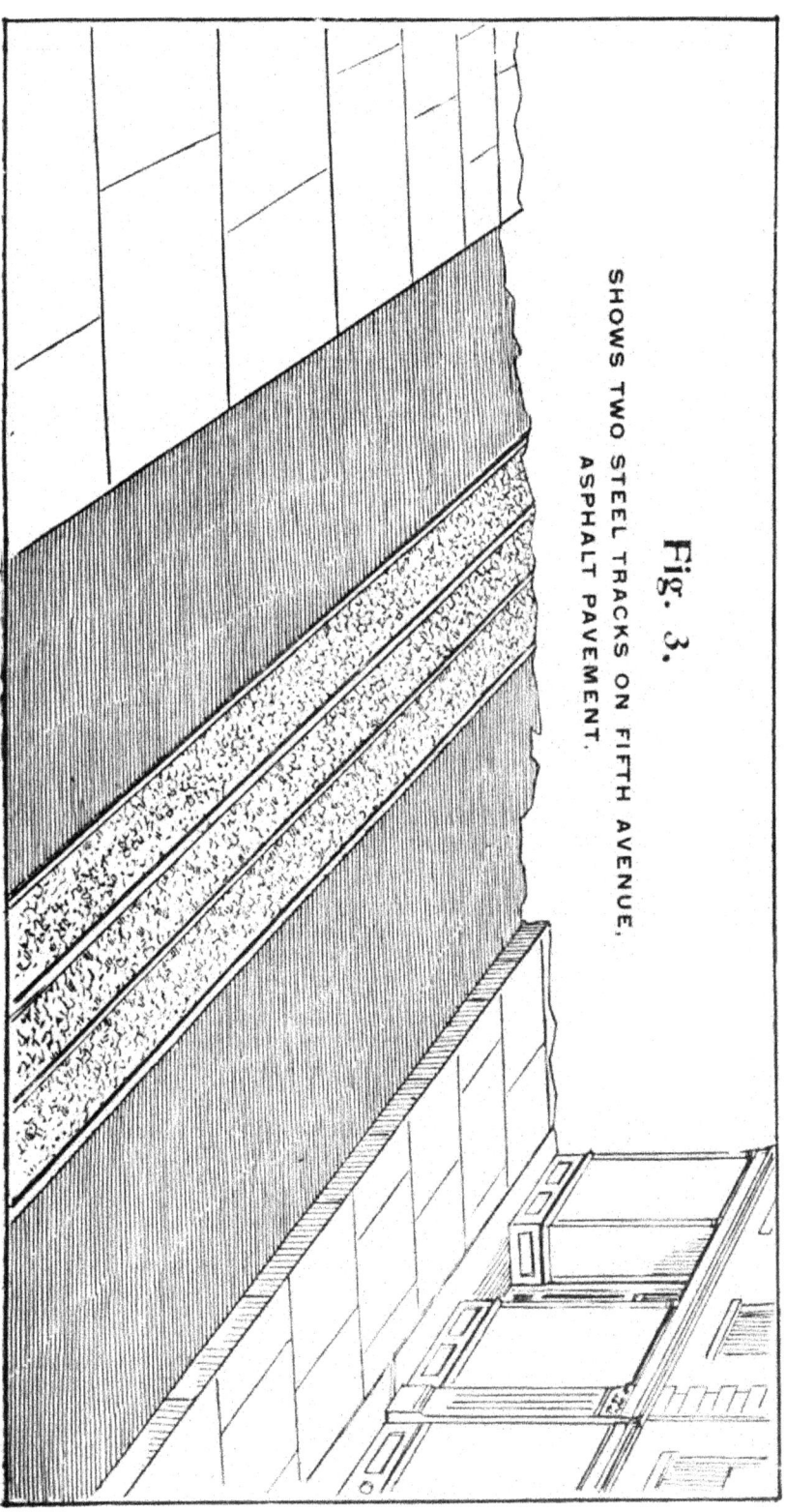

Fig. 3.

SHOWS TWO STEEL TRACKS ON FIFTH AVENUE, ASPHALT PAVEMENT.

pared with electric or cable roads, steam motor roads have decided advantages. They involve no investment or cost of maintenance for battery of boilers, stationary engines, power house, dynamos, overhead poles and wires or underground conduits, cables, right of way, etc., etc.

The experiment of steam traction only involves the cost of one motor (which can be sold at any time second-hand), and two or more passenger cars. Passengers could be carried over such a road at a cost of half a cent per mile, in connection with freight and express at half the rates now charged by the railroads. Athough we do not advocate the leasing of the public highways generally, yet it might be well in some sections to give contractors a twenty years' lease on condition that they build and maintain such steel-track roads, at the same time giving the people along the line of the road a stipulated price for passenger and freight service, always reserving the right of the people to use the steel-track road. It is believed that with such an offer contractors would occupy all the principal roads in every State. With a careful contract such service might be leased at profit by both parties.

WILL SEAT
25 PEOPLE.

COST FROM
$400 TO $600.

Since writing the above we have received the following from one of the largest locomotive works in the country, in reply to a letter of inquiry:

"We would figure that to draw one light car, say with twenty-five passengers, weight of car and passengers about 5 tons, that a locomotive with cylinders 5 inches diameter by 10 inches stroke and weighing 4 or 5 tons would be abundantly powerful; and if it were desired to carry an additional small freight car, or to haul a heavier load of passengers, you might find cylinders as large as 7x12 inches, and weighing in running order 8 to 10 tons, desirable

"The height of locomotives of this class could easily be made not to exceed 8 feet, and in no case would it need to exceed 9 feet or 9½ feet.

"The cost of locomotives of this class might run from say $1,500 up to $2,500, more or less, according to details of design.

"We figure that for a fifty-mile run you might need say 300 to 500 pounds of coal, and the daily wages of a suitable engineer would probably be $2.50 to $3 00. We would figure that 50 cents a day ought to cover cost of oil and repairs.

"In the matter of speed there will be no difficulty whatever, as these little locomotives can run a great deal faster than they will be permitted to run on ordinary roads. We would figure on say ten to twenty-five miles per hour.

"Of course, the lighter the grades are the smaller the locomotive required to haul the given train. There is nothing impracticable in using motors of this class on grades as steep as 8 or 10 degrees.

"It occurs to us to add that you might find—in some localities at least—that it would be the simplest and best arrangement to use a steam tram-car, carrying the fuel and water and passengers all in one machine. We could get up a very serviceable arrangement of this kind, which would carry a sufficient number of passengers and make a mileage of 100 or 200 miles a day, at a cost of something like $6.00 to $8.00 per day for operating, or say 3¼ to 4 cents per car mile, burning about three-quarters of a ton of coal and employing one motorman and one conductor."

STATE AID FOR GOOD ROADS.

Does any intelligent person doubt that it is a good State policy for the State to appropriate money to aid in the building of good roads? If so, a few facts will be in point. France has the best roads of any country in the world. It has long been the settled policy of France and other European countries, as well as England, to aid in building country roads. The French Republic has more than 130,000 miles of macadamized roads. The Government spends $18,000,000 per year to keep these roads in repair—saying nothing of the original cost of construction—and finds it a good investment. At the close of the German war the world stood aghast at the money indemnity demanded by the victorious Germans—a thousand million of dollars! But France offered the loan to her peasant farmers and it was subscribed three times over in a day. The peasantry of France had enjoyed the protection of the government in good roads, and, as a consequence, were able and glad to take this vast loan from the government at a very low rate of interest to find a paying place for their hoarded wealth.

Massachusetts statesmen are proverbially money wise. In 1893, after an exhaustive report to the Legislature, an appropriation was made of $300,000; the next year $400,000, and this year $800,000. And almost every town in the State has petitioned the commission for State aid for town roads, although the county pays twenty-five per cent. of the cost. State aid is the rule in all the New England States.

Among the States Kentucky took an early lead in co-operation with counties, municipalities and private capitalists in the construction of turnpike roads. The State contributed about $1,750,000 to seven of the leading thoroughfares, covering 640 miles, and this is only a portion of its total expenditure. In the years 1837 and 1839 the State had in its employment an engineering corps, principally engaged in road works, costing an aggregate of $31 675 per year. These improvements covered only a trifling percentage of the total mileage of roads in the State, but they have been of such value as to make the State conspicuously prosperous for the last half century. More recently the co-operation of counties with the local authorities and property holders, upon the basis of a contribution of $1,000 per mile by the county to every mile of good road built, is securing a wide extension of road improvement.

The State of Ohio, observing the advantages of good roads to her neighbor, has followed with very extensive road improvement. About one-eighth of the total mileage of roads in the State has been improved.

The State of Massachusetts has taken upon itself the entire burden of building the principal roads throughout the commonwealth, though it ultimately requires the counties to pay one-fourth of the cost.

Connecticut has taken up the co-operative method upon the scale of an equal distribution between the State, county and the district called town (which is equivalent to a township in the Western States.)

Many other Southern States have endorsed the policy of State aid. New Jersey, after appropriating $350,000, voices her sentiment in the annual message for 1896 by Governor Werts, as follows: " The question as to the good policy of the State aiding to build good roads has passed beyond the stage of discussion by an overwhelming decision in the affirmative," and he continues:

"As all progress should be along the best lines, and as some of the best authorities predict, the coming highways will be of steel. It would be well for New Jersey, as she is the pioneer in State aid road improvement, to take the lead in inaugurating a system of steel roads, and thus ascertain by actual experience whether it is the most efficient and economical highway. The claims as presented are that the average cost of a macadam roadbed sixteen feet wide is about $7,000 per mile. The cost of a double steel track highway, sixteen feet wide, filled in between with broken stone, macadam size, is about $6,000 per mile. The cost of a rural one-track road, $2,000 per mile. The rails to be made of steel the thickness of ordinary boiler plate, and to be formed in the shape of a gutter, five inches wide with a square perpendicular shoulder half an inch high, then an angle of one inch outward, slightly raised. This forms a conduit for the water and makes it easy for the wheels to enter or leave the track. The advantages of steel rails are, first, longer wearing qualities than stone; second, one horse will draw on a steel track twenty times as much as on a dirt road, and five times as much as on macadam. We would therefore recommend that legislation be enacted and appropriations be made by which the Commissioner should be empowered to authorize the construction of some experimental steel track roads."

As to the financial policy of the State aiding in building good roads, there seems to be little question. But, strange to say, there are some individuals in every community, even among farmers, who are opposed to any State aid legislation. To such the following extracts from the New Jersey Commissioner's report is respectfully presented. The Commissioner says when the first State aid bill was presented it met most opposition from farmers, but since the roads have been built these same men have become the warmest advocates of the system. Why not, when their farms, heretofore a drug on the market, upon the completion of improved roads, became salable at an advance in price of from $10 to $50 per acre?

As many State legislations are now in session, we herewith present a skeleton bill, drawn by the writer, which has once passed the Assembly of the State of New York, and is now pending in the Senate, hoping that it may become the basis of a bill for some other State that has not yet moved in the direction of direct aid in building good roads. It will be noticed that the proposed bill leaves it for the people to decide the kind of road—in fact, the majority of the people decide whether they will avail themselves of such a law or not.

AN ACT
To Promote the Building of Good Roads.

SECTION 1.—The State Engineer shall have charge of the highways and roads of the State, etc., etc.

SEC. 2.—It shall be his duty to hold public meetings in every county of the State, once in every year, for the purpose of awakening public sentiment upon the subject of good roads, etc., etc.

SEC. 3.—Improved roads shall be known as "improved county roads" and "improved town roads." County roads shall run through a county, as near as may be, north and south and east and west, on main roads leading to and from markets, etc., etc.

SEC. 4.—Whenever a petition shall be presented to the State Engineer for an "improved county road," signed by a majority of the owners of the lands bordering on such road, or by a written request of a majority of the Board of Supervisors, or by a majority vote at a meeting of the Board, he shall inspect such proposed road and make a written report, with a recommendation of the kind of road best adapted, etc., etc., with an estimate of the cost, and shall mail fifty copies of such report to the supervisors of each town.

SEC. 5.—It shall be the duty of the supervisor of each town to call a special meeting of the voters to vote upon the question of an "improved county road," etc., etc.; and if a majority of the voters of a town shall sign a petition in favor of such improved county road, and the kind of road, etc., etc., then, in that case, no town meeting shall be held; * * * and if a majority of the voters of the town, either by vote or petition, vote in favor of such county road, he shall proceed to let the contract for building. * * *

SEC. 6.—The cost of building a "county road," as provided in this act, shall be paid one-half by the State, and one-quarter by the counties through which the road runs, begins and terminates, and one-quarter of the cost of such road shall be paid by the towns where such road begins and ends and through which it runs. * * * The Board of Supervisors of any county, or Town Board, may borrow money from time to time to pay their proportion of the cost of such road; and such county and town may issue county or town bonds for a term not exceeding twenty years, at a rate of interest, etc., etc. * * * The State Engineer shall appoint a competent local overseer of such road upon the written recommendation of a majority of the Board of Supervisors. His salary shall be three dollars per day for actual service, and traveling expenses. The expense of maintaining an "improved county road" shall be a county charge.

SEC. 7.—Whenever an "improved town road" is asked for by a petition signed by a majority of the owners of land bordering or abutting on the proposed road, the State Engineer shall make an outline map, and the Supervisor of such town shall proceed to take the vote by petition or town meeting, and if a majority favor an "improved town road" it shall be the duty of the town to build the same within one year, on the plans recommended by the State Engineer, and the town shall, by tax or otherwise, raise the money necessary to pay its share of the expense as herein provided, and when completed, the road within each town shall be maintained by such town. The cost of building such town roads shall be paid one-half by the State and one-half by the town. In case any individual or association shall, in writing, notify the State Engineer and Surveyor that he or they will pay one-half the expense of building an improved highway, such individual or association may build such road under the supervision of the State Engineer, and the State shall pay one-half the cost thereof, etc., etc.

There are many localities, particularly in the neighborhood of New York, that have not waited for State aid, but have bonded their towns large amounts to build good roads, which they are now enjoying, and the rise in the market value of their lands has more than paid the debt, although the bonds run for twenty years. Many counties in other States have been bonded to build good roads, notably Shelby County in Ohio, $2,500,000.

While the States ought to lead in this work, counties should not wait, but should move at once. County bonds are considered the best of securities. These bonds could run twenty years and at the end of that time be refunded like railroad bonds, and this generation enjoy the luxury and profit of good roads. Provision also could be made in any future State aid bills to give such counties their equitable portion.

INTERVIEW WITH EX-SENATOR O'DONNELL.

REPORTER: Senator, you have seen many years of public service in the State and have the reputation of being posted on taxation and economic questions. Will you give the *Republican* something for its readers?

ANSWER: Yes, I have given the best years of my life to the State. I was Clerk of the Assembly two terms, Member one year, Senator four years, United States Supervisor of Internal Revenue for all of the State west of Albany four years, and State Railroad Commissioner four years. The Commission law devolved the duty of naming one Commissioner, to represent the mercantile and farming interests, upon the Chamber of Commerce of the State of New York, the Board of Trade and Transportation and the National Anti-Monopoly League, and I was their choice. State Railroad Commissions are good in their place, but generally the railroads in fact name the Commissioners. The Inter-State Commission was an afterthought, to do what State Commissions were powerless to do; but here the railroads won again, for a majority of the first Inter-State Commission were former railroad attorneys. The simple facts are that the railroads of the United States are more powerful than the Government. They rule by electing representatives friendly to their interests generally, and, when occasion requires, they make friends in their own peculiar way.

REPORTER: Senator, to what do you attribute the hard times we are passing through as a nation?

ANSWER: It is not the currency question, nor is it the tariff primarily, although both of these have their influence, but the one procuring cause, and that lays at the foundation of all our financial difficulties, is the decline in agriculture. Washington, in his farewell address to his countrymen, warned them in prophetic language that "upon the prosperity of agriculture depended the prosperity of the nation." This sentiment was repeated by Jefferson, Madison and all the early statesmen. What do we see to-day—agriculture neglected, farms mortgaged, value of land declined as never before in the history of the nation, with absolutely no hope of any permanent improvement of land values in the future. It becomes the statesman to pause and enquire what is the foundation trouble. Take the State of New York for illustration. It is a historical fact that the decline in farm land values began upon the completion of the great trunk railroads running through the State to New York City. These roads at first were aided financially by the State. Over $10,000,000 at different times have been donated by the State. Then came the era of bonding roads and creating what is called "Fixed Charges." These generally are either the actual cost of the road or represents water. As the result, the bonded debt and capital represents more than fifty per cent. of water, and upon this vast aggregation shippers are forced to pay interest in the shape of large dividends. The profits of farming are so small that the increased freight charges to pay interest on this inflated capital upon the products of the farm sent to market absorbs all the profits. In short, this is the trouble with farmers of the State and Nation. You may change the tariff and the currency, one or both, and the farmer will receive little or no relief. What he wants is an open field and "half the road," and he will take care of himself; but tie him down and handicap him in freight and transportation of his farm products to and from markets, and he sinks into despondency. It's no use to rehearse this old story. The question now is, what are you going to do about it? It's a live question, and must be met. Does any sane man suppose that this country can go on another ten years with farm lands declining in price as they have in the past ten years? Why, sir, the farmers in that time will be paupers, and this will be a nation of bankrupts, for there is no truer economic fact that as the farmer prospers the nation prospers. Politicians tell us a change of administration will change the times. Not so. You must again build anew the foundation before there will be a permanent change for the better. It is agriculture that needs the protecting hand of government, or if the ghost of a "paternal government" is raised by the corporations who have fattened on protection call it justice to farmers and return to them a part of what has been robbed from them by corporations created by the government, in its *paternal* care.

REPORTER: Senator, what is the remedy for the hard times farmers suffer from?

ANSWER: Some statesmen who have fully grasped the situation have proposed that the Government take all the railroads at a fair valuation, and run them in the interests of the people. They point to the vast operations of the Post-office department to show that the Government can run railroads, and in addition cite the facts that the governments run successfully the principal railroads on the Continent. These statesmen see clearly the wrong done the farmers by the railroads of the nation. But the task of buying and managing the railroads of the nation—even if legislation could be had—is herculean, and the boldest shrink back at the task. But the remedy is at hand directly on the line of governmental ownership. For the people are the Government. And the people own the 1,500,000 miles of the public roads of the nation. The steel track highway is the boon presented to the people—the great deliverer of the people. Although the trolley roads have begun the work of capturing the public highways there is time to call a halt. Every taxpayer in America should set his face as a flint against every proposition to capture the public highways. There should be a constitutional amendment adopted in every State forever prohibiting the sale of the public highways to transportation corporations. It might be permissible in some cases to lease a part for a limited term, but in such cases it should be only to get a good road for the use of the people entirely free.

REPORTER: Senator, how would you have steel track roads built?

ANSWER: Let the people, by towns or counties, as the case may be, issue twenty-year bonds at a low rate of interest and build their roads at once. Every State is now willing to aid, more or less. When these aid bills are passed, let such counties as have been bonded see to it that they are equitably provided for. The building of steel track roads in one county will incite to the building in adjoining counties, and the whole movement will help push on State aid bills. Here you have a system of government—the people—owning the steel roads of the nation in spite of corporations. It's the simplest and cheapest form of railroad ownership ever devised. No buying of right of way, no bonuses, no water, and, best of all, the people the owners. And it would be a sovereign remedy; for it is a wonderful fact that slow freight and passengers can be carried by horse or motor power on such a road at half the present prices charged. Along the line of such roads there would be hundreds of manufacturers of wares and merchandise. Every little water power of three or four horse-power would be utilized. Roads would run from these main lines to reach these water powers.

REPORTER: Senator, are steel tracks all alike and uniform, with no change of gauge?

ANSWER: Yes. This is the beauty of the system. No matter whether the steel track be heavy or light, on a main or cross road, the width of the steel track is the same, and a man with his team or on a motor, when the track is once laid, can travel all over the United States without "change of car or baggage."

REPORTER: Senator, how will the steel track system affect the steel trade?

ANSWER: It will create an enormous home demand. The steel

and coal interests will experience a great boom. It is doubtful if, in five years, all the present rolling-mills in the country will be able to supply the demand for rails. If I had money to invest I would put it in Blast Furnaces, Rolling-mills, Ore Beds and Steel Track Highways. Such stocks will inevitably go up and be of permanent value. The general business of the country will receive an impetus from steel track roads. It will not be a short, spasmodic boom. On the contrary, it will be permanent and lasting, reaching into every department of business. The grain growers of the West will be perfectly independent of trusts and combines. Now the farmers' grain is stored in corporation elevators. When they are ready, they fix the price, and, however low, farmers have no alternative but to sell. The price of grain is not fixed upon the law of supply and demand, but upon betting contracts for future delivery. These gamblers have the whip-row so long as the grain is held by elevator monopolies. But give the farmers steel wagon roads and the situation is at once changed. With the grain in their own hands ready to deliver, they can largely fix the price. The buyer in such a case will come to the farmer and ask him his price, instead of the farmer going to the buyer—with the cards all stacked against him—to sell. When the farmer prospers manufactures will boom, for a home market will spring up where now is all dearth and want. Give the farmers the boon of steel track highways, and the old spirit of independence and self-reliance will spring up as by magic, and we shall again be a nation of freemen instead of industrial slaves.

REPORTER: Senator, do you anticipate that travel will be increased as well as freight by a steel track road?

ANSWER: Travel increased! Why, sir, with a system of steel tracks from the West to the East there would be a reunion of families long parted by a distance so expensive as to forbid even the parent or child from meeting for a score of years. Talking from the East, in reply to the badinage or gibes of the West that we of the East are "slow and conservative" and lack the vim and push of the West, we remind them that it was the flower of our sons and daughters that peopled the West. Give us steel roads and the old home will be brightened by the annual visits of these our loved ones. Travel, sir! Why, there would be no end of the train of visitors to and from the West, and instead of the cost of travel in the way of railroad fares being sent to Europe to pay interest on railroad bonds, it would be distributed more or less along the lines of these steel track highways. Some years ago there was an era of plank roads in the State of New York. Along these roads lands at once increased in value and travel increased immensely. These roads were soon worn out, but in their short lifetime of ten years country hotels sprung up on an average of one in ten miles to accommodate the traveling public. To-day they are deserted. Give such roads steel tracks and land will advance in price, and the increased travel will reopen all these old-time hostelries.

REPORTER: Senator, do you expect the people or contractors will build these roads?

ANSWER: Both will undoubtedly build roads. Contractors who are fair and honest can build roads at a fair profit and for less than a community. It takes tools and a plant, with some considerable capital, to build. The company gives responsible contractors territory and refers all letters relating to the system to them. In the next edition of the "Hand Book," to be published next month, the company will publish the names of contractors and the territory assigned to them. Where there are no contractors owning territory, the company deals with individuals, towns or counties, but encourages the formation of new contracting companies, as it is more convenient to deal with such. The company has already applications from States all over the Union, but as the parent company has been organized but a few weeks it has waited in arranging territory, but more particularly to get its "Hand Book" out, to answer questions which would take a large clerical force to otherwise answer. The press, with not a single exception, have endorsed the steel track highway system, and there is no doubt that it is the coming road of America.

QUESTIONS ANSWERED.

We are continually receiving letters of inquiry and answer as follows:

QUESTION: How high do you make the grade before you lay the cross ties?

ANSWER: The grade depends on the conditions and facilities for draining the water off at the side of the road. The grade must conform somewhat to this factor. Upon an ordinary level road, where the drainage is easy, a side drain ten inches deep will be sufficient. This, with the road superstructure, will make the drain 18 or 20 inches deep from the center of the road. If proper drainage can be obtained it is better not to disturb the surface of the old road bed. Where there are fills to be made on the grade, gravel, sand or clay is preferable unless the rest of the road is stoned—the road surface, as near as may be, should be alike. The road bed, however, should have enough soft material on top to make it easy to level the ties. Upon a sand road no drainage is necessary, and when the surface is leveled the ties can be laid, followed by the stringers and steel rails. A road on sand can be built very cheaply and quickly. If there is a clay bed anywhere on the line of the road, a first-class road bed between the rails and at the side can be made by alternating layers of sand and clay, to be finished by rolling with a five-ton roller.

QUESTION: What are the dimensions of the ties and longitudinal timbers?

ANSWER: On a sand road the ties may be 2x8 inch plank, laid three feet apart from center to center. On a hard dirt road the ties may be 3x6, laid flatwise. These may be alternated with heavier ties in wet spots. In wet, swampy land the tie should have a broad, flat surface, about 3x8 inches, or even wider. In wet places ties of any proper size may be hewed flat on one side, in fact, such ties may be used anywhere; the only question on a hard road bed is whether it is not cheaper to have the tie flat on both sides to facilitate leveling up. The longitudinal timbers on which the track is laid are preferably four inches thick and five and a half inches wide, which tills the rail. A thin-sawed wedge piece is tacked on the longitudinal timber under each spike between the rail and the timber to make it solid. The longitudinal timbers should be spiked to the ties from the edge diagonally, which is stronger than if driven straight down, as is the case when the rails are spiked to the timber (see cut No. —). The ties should be six or seven feet long, so as to project beyond the edge of the rail. On a sandy or soft soil longer ties are better by supporting a load in turning out. It will take 1,760 ties, laid three feet apart, and 2,640 two feet apart. The distance apart is regulated by the character of the road bed and size of the tie used. On a soft, wet road bed the ties should be closer together.

QUESTION: How close together do the side conduits from the tracks want to be, and how are they built?

ANSWER: The side conduits are among the most important features of the steel track system. No matter what kind of an improved road is built, unless the rainfall be taken care of the road will be ruined. The gutter track takes care of the water at once, but there must be an outlet or the track will be flooded. In general, a side conduit must be made as often and where there is a depression in the road. This side conduit may be made of pieces of the steel track or by a paved gutter. The latter is cheaper and just as good. When made of cobble stone the earth must usually be excavated below the frost and about two feet wide, and then filled in with loose stone or sand, and a nice paved gutter, a little descending and a little larger capacity than the track. On a single-track road a gutter may be made to each side of the road, or a gutter may cross between the tracks opposite the side conduit, paved as in the first instance. In a double-track road this plan must be adopted. Care must be taken that the gutter is made solid where the horses tread—in fact, the gutter and every other part of the road should be built in a good and workmanlike manner. It does not pay to slight the work, notwithstanding a steel track road never ruts or becomes uneven like a dirt or macadam road. At the time of ordering rails the number of side conduits should be determined, so the rails can have requisite openings, which are simply taking the flange off six inches. The stone pavement must extend up under this by cutting into the stringer, so that there will be no water set back to rot the stringer. A little asphalt may be profitably used at the joint. The pavement must in any event have sufficient pitch so that the water will run off quickly. Where the ends of the rail meet, a galvanized thin sheet of iron, eight inches wide, wraps on the top and sides of the timber, so that no water will rot the timber. It will be noticed that there are no spike heads on the track.

QUESTION: How about a loaded team turning out? Won't the gutter track hinder?

ANSWER: The track is only half an inch high and rounded at that. In the cities the car tracks are three inches wide and one inch high, against five inches wide and half an inch high. City truckmen, with two or three tons to a load, always try to get on to the car tracks because of the easy trackage, and they are whistled off every few rods by the car drivers. If on these narrow, one-inch tracks there is no difficulty in turning out, surely there will be none on a five-inch track half an inch high.

QUESTION: What is the cost of a good, first-class steel track road?

ANSWER: It is difficult to answer definitely the question of cost until one knows the character of the road bed, grade, proximity of materials, etc. In general it may be said that the cost of a steel track road will be cheaper than a first-class macadam road. It is understood in estimates that towns or counties, as the case may be, build their own bridges and approaches. Under favorable circumstances a single-track road with the track at one side of the grade, 13 feet wide, can be built from $3,000 to $5,000 per mile. A double-track thoroughfare, where there is heavy travel, built as described above on a fair grade will cost from $6,500 to $8,000 per mile. The reasons in general why such a road is cheaper than a macadam road are, that on a macadam or Telford, first, there must be a channel or canal made for the road bed. This involves the removal of a large quantity of earth, which is very expensive. Then, each successive layer of stone must be rolled and finally finished with fine stone smooth for the trackage of the wheels. The process is very slow and tedious and must be done in dry weather. There is comparatively little work in preparing the road bed for a steel track. After the ties and tracks are laid the material can be hauled on the steel tracks, the road bed being packed at the same time by the tread of horses and finally by our improved dumpers, scrapers and rollers, the whole road is finished; the improved machinery, saving at least the work of twenty men, who must be employed in building a macadam. On a first-class macadam road the saving above given will pay for the steel track.

QUESTION: What thickness do you recommend for the stone filling or crust of the road?

ANSWER: The crust of the road, generally speaking, need not be more than half as thick as a macadam road, because the ties, stringers and steel track form a crib to hold the material in place, which, as every engineer knows, is a very important factor. If a six-inch tie is used and a four-inch stringer, with two inches allowance for the crown of the road, we have twelve inches from the bottom of the tie to the crown to fill. If the tie be only four inches, which is heavy enough for an ordinary hard road, the fill will be ten inches. If the tie be only a two-inch plank, which is heavy enough on sand, the fill will be only eight inches. The bottom between the ties and two inches over may be filled with stone, gravel, sand, clay, or any material that will pack solid. After the bottom is rolled or compacted together solid by teams driving over it (which is the very best way to make a solid road), the top is finished with crushed stone, about two inches size, and a surface of stone screenings, very coarse sand or blue clay, when it is thoroughly rolled by a five-ton roller. The grade should be finished in the same manner and thoroughly rolled. It will be noted that the steel track is laid before any filling is done. If there is a sand or gravel bed, stone, slate or blue clay along the line of the route this material can be drawn twenty miles on the steel track cheaper than on one mile on a dirt road, and the tread of the horse, going back and forth (for but one horse will be needed), will make the road bed much better than if rolled. Sand or gravel at the bottom, with a blue clay surface, mixed in layers and finished with a coat of sand, thoroughly rolled, will make the road bed between the ties as hard as tile and a good hard road on the grade. A double-track road is built in the same manner, the grade between the ties being filled a little crowning, so that the water will run off from the center to either track. A six or eight-foot rural, or cross road, is built in the same manner, and where material is at hand can be built very much cheaper.

QUESTION; How about the durability of the ties and timber?

ANSWER: Much depends upon the kind of timber. In different localities different timber will be used. The average life of a railroad tie is about ten years under favorable circumstances. But these ties are exposed to the most destructive influences. They are alternately exposed to wet and dry weather and the sun, thus disintegrating the fibre of the wood, which in turn is filled with the rainfall. The ties and timber in the steel track system are housed by a cover, and are never wet or exposed to the sun. This protection will more than double the life of timber. I asked one of our State Forest Commissioners how long timber would last covered up from the air, sun and rain. He replied, "If you cover up timber as proposed it will last forever, or, rather, it will last anywhere from twenty to forty years. I have just been taking up some old fence on my farm, built forty years ago, and while many of the posts are rotted off above ground, the bottoms are as solid as when first put in the ground."

QUESTION: Do you think the country can stand the taxes to build your system of steel track roads?

ANSWER: Stand the taxes! Why, sir, the building of steel track roads means a lessening of taxes. Take the State of New York, for an example—and almost all the States have copied our bad-road laws. By request of Gov. Flower, in 1893, the county clerks made returns of all the road taxes assessed in the country districts of the State, not including cities and villages. It was the first tabulated return ever made to the State, and it amounted to the enormous sum of $3,500,000. Add the road tax of villages, which properly belongs to the country districts, and the annual road tax is over $4,000,000. One-half or more of this vast sum is paid in cash, the commutation rate being $1 per day of assessment. This annual tax, if all paid in cash, would pay the interest at 3 per cent.—the price of good State and city bonds—on over $133,000,000. What does the State get for this enormous annual road tax? Comparatively nothing. We have justly the reputation of having the poorest roads of any State in the Union. Is any man so blind that he cannot see that it would be a great financial operation for the State to issue $100,-000,000 twenty year bonds to build good roads, and take this road tax to pay the annual interest and create a sinking fund to pay the bonds? This tax has grown up gradually in the State until last year in the State outside the cities, including villages, the highway tax is fairly estimated at $4,500,000, over one-half of this sum being paid in cash. (See returns from counties made in 1893.) This tax amounts to more than the State tax for State prisons, civil and criminal courts, penitentiaries, insane, hospitals and charitable institutions. It is more than the entire tax for schools, academies and institutions of learning. The annual road tax of the State, if all paid in cash and honestly spent, since the Erie Canal was first projected, would have built the entire canal with its enlargement, and the balance would pay for a steel track highway over every principal road in the State. For all this enormous tax what have we got? While speaking at the Atlanta Road Parliament I was interrupted by a delegate from Kansas with the statement: "But New York has the reputation of having the poorest roads of any State in the Union," and I couldn't deny the statement, for it is conceded to be a fact by all who have studied the road question? Talk of the State being able to stand the tax! The State absolutely throws away annually more than the most ardent good road advocate would ask it to appropriate. Probably there was never in the history of government such a prodigal *criminal* waste of public money as the highway tax in the State of New York. A similar examination in many other States will show similar waste and negligence. It's high time the people woke up to this subject. If they understand that an honest application of the money for roads now raised annually will pay the interest on twenty-year road bonds and the principal at maturity, how long will it be before a tide of public sentiment will set in for this great State and national reform culminating in steel track roads on every public highway in the United States? In short, the taxes we now pay for bad roads will build good roads by paying the interest on long time bonds, besides taking care of good roads and creating a sinking fund to pay the bonds at maturity. In suburban villages the steel track laid through a principal street or avenue, forming a connection with a railroad or water communication, would advertise such a suburb more than many columns of newspaper ads., and would in the early stages of development take the place of sidewalk and gutters, as the centre of the track of itself would be a good walk, beside its use for carriages wagons and bicycles.

QUESTION : Are you the inventor of the steel track for highways?

ANSWER : Yes, but I am not a contractor or road builder. A patent was issued some three months ago. I had sold it before issued, but I have an interest in its success. I am general manager of the O'Donnell Steel Track Company. My attention was first called to a steel track for the wheels on wagons by reading a newspaper paragraph, that a horse would draw twenty times as much on a steel track as on a dirt road. Bosh! I exclaimed; another newspaper yarn. It cannot be possible that all the teaming and hauling of freight and produce in this country for the past hundred years has been done at such an economic loss. I sent the clipping to the *Scientific American* and asked if the statement was true. I received an affirmative answer, with plenty of engineering authority. This led to a thorough investigation, and I found that a steel track road could be built as cheap or cheaper than a good macadam or telford road, and that a team could draw five times as much on a steel track as on a macadam, and that the cost of repairs after built would not be a quarter as much as a stone road.

QUESTION : Will steel tracks be used in cities ?

ANSWER : Every argument for steel track country roads applies with tenfold force to city traffic. A horse will draw twenty times as much on a steel track as on a dirt road, and five times as much as on a smooth road, and ten times as much as on a stone pavement; but call it half this, or five times as much, and the saving of expense in haulage runs up to an enormous amount. Add to this the saving of time in moving freight—for a horse will draw his load in half the time on such a track. Then the saving of horses—for one horse will take the place of two or three now used for heavy loads. Then comes the wear and tear of wagons and carts, etc. It is believed with a steel track the saving in broken carts, wagons, harness, etc., would go far to pay for steel tracks. The savings in the matter of pavements would be immense, taking into consideration the life of steel tracks. The cost of repairs to city pavements amounts to a very large sum, a large part of which would be saved by steel tracks, for it is the pounding of the wheels that causes the ruts which obstruct travel and spoil the roadway for other vehicles. As it is now, on every street car railroad there is a continuous rut from four inches to twenty, caused by loaded vehicles seeking even one track for their tired horses. All this would be done away with by the one wide track, which provides for different width axles. The proposed street track is much superior to the present style of street railroad tracks, and all such companies will soon use the steel highway both for street and trolley cars. The steel track has no headers, and is much superior to the "strap" or "I" rail. The wide track would not interfere, while it would entirely prevent the rutting wh'ch destroys the streets and causes large expense to the street railroads. The track proposed for Fifth avenue would make this street, when supplemented with asphalt pavement, the model drive of the world. With the near approach of the Horseless motors it will be only a short time when horses will be dispensed with on this avenue, and in place of ponderous carriages light fairy vehicles will be seen everywhere where the steel track is laid.

QUESTION : Do you expect the people or contractors will build these roads ?

ANSWER : Both will undoubtedly build roads. Contractors who are fair and honest can build roads at a fair profit less than a community. It takes tools and a plant, with some considerable capital, to build. The Company gives responsible contractors territory, and refers all letters relating to the system to them. In the next edition of the Handbook next month the Company will publish the names of contractors and the territory assigned to them. Where there are no contractors owning territory, it deals with individuals, towns or counties, but encourages the forming of road contracting companies, as it is more convenient to deal with such. The Company has already applications from States all over the Union, but as the present Company has been organized but a few weeks, it has waited in assigning territory, but more particularly to get its Handbook out to answer questions, which would take a large clerical force to otherwise answer. The press, with not a single exception, have endorsed the steel track highway system, and there is no doubt that it is the coming road of America.

HOW TO BUILD GOOD ROADS.

JUDGE THAYER, AT THE NATIONAL ROAD CONVENTION.

* * * "I do not believe it is practical to make good roads in conformity with the advanced ideas of the day without a change of base. For three-quarters of a century we have been trying to make roads that way, and each year the mud gets deeper.

"Not only must there be a radical change in the manner of paying road taxes, but the money thus paid must be expended in a different way. The local method of building roads must in a great measure be abandoned. The next generation must be asked to help bear the expense of building the roads which the next generation will enjoy. To do this the road taxes need not be increased, but use the taxes to pay the interest on loans for money advanced to build good roads economically and on an extensive scale.

"Construct roads on the same plan whereby the great enterprises of this land have been built up. If it is thought the best policy to limit road-building to a county and not make the State the chief factor, provide that all the road taxes shall be paid into one treasury and, instead of being used in the repairing of the roads already built, devote the larger portion to building permanent roads and the rest to repairs. If there should be a prejudice in any county against borrowing money on long time bonds at low rates of interest and spending the money as rapidly as it can be done to advantage, and using the taxes to pay the interest and creating a sinking fund to pay the principal when due, then adopt a plan for building, with the annual taxes, a certain number of miles of good road every year. Different communities will have different views as to which policy is best to pursue. But it is well enough to bear in mind that the larger number of the great improvements in this country have been brought about on borrowed money. One man never undertakes to build a railroad. For one man, or even one community, to undertake to build so much of a railroad as runs through his or its school district would be a slow method of building trunk lines of railroads. It might be done that way in time, but railroads have not been constructed in that way. The vast railroad system of this country is the work of the ablest financial geniuses, the best skilled engineers, the most successful business men the century has produced, and I believe that to-day, without loans on bonds, there would be less than 20,000 miles, of road where there are 200,000 miles. Other great industries conducted on a colossal scale, and which are the pride and boast of the nation, owe their success to a combination of purses advancing money to be repaid in the future.

"The rich treasures of the earth are brought from great depths to minister to the comforts of mankind by the multitude putting their money and their brains together and making investments which no one individual cares to do. The immense resources of this country have been and are being developed by the co-operation of men using their money jointly for common purposes. I believe that the practical plan of road-making is to follow in the tracks of men who have taken giant enterprises out the line of experiments and made them giant successes. So it is no unexplored field I take the public into when I ask it to enter upon a system of road-making that shall equal any undertaking in which the country has ever engaged, not excepting the building of nearly 200,000 miles of railroad." * * *

GOOD ROADS IN NEW JERSEY.

DR. CHAUNCEY B. RIPLEY.

"'How good roads were obtained in New Jersey,' is the subject, Mr. Chairman, which you have assigned to me.

"I have intimated already the bad management and deplorable condition of our public roads for long years before we entered upon the present era of reform and improvement. Once New Jersey was proverbial for red mud and bad roads. Now Union County has the best system of public roads in the United States. We have forty miles of telford pavement. It is in two continuous lines, crossing the county; one road extending from the city of Elizabeth, the extreme eastern boundary of the county of Union, to the city of Plainfield, the extreme western boundary; the other road extending from the city of Rahway, the southern boundary of the county, to Summit, the most northern boundary. These roads are in contact with every other city and township of the county, intersecting at Westfield, the center. The cost of the roads was $400,000, or $10,000 per mile. They were constructed under an act of the New Jersey Legislature passed in the year 1889. They have proved a success to this extent: there is not a citizen of the county of Union known who does not regard the improvement as the most important and satisfactory of any ever accomplished in the county. The cost of the roads did not exceed 1 per cent. on the assessed valuation of the taxable property of the county. The advance of values in the real estate of the county since the roads were built is conceded to be from 5 to 25 per cent., an average of about 15 per cent.—that is, 5 per cent. on farming lands and 25 per cent. on other lands. Besides, the saving of wear and tear to vehicles and draft animals of every sort and kind, almost, if not quite, compensates for the annual assessments on account of the improvement. Moreover, the convenience and comfort of these improved roads to citizens generally preclude the idea of ever doing without them again. As well might one advocate to the people of Union County that they return to the tinder box and flintlocks as to mud roads. Good public roads, the best practicable and obtainable, are alone consistent with the progress and civilization of the present age. It is a disgrace to a civilization like ours that in a community like that of Union County, N. J., and Westchester County, N. Y., and in other communities suburban to the national metropolis of this great country, and in this wonderful nineteenth century, the people have not public roads over which they can travel with ease and comfort at every season of the year. It is hardly less a disgrace to the civilization of the age that the citizens of the metropolis of Illinois and of Omaha, Nebraska, and most of the great cities of the West, as Senator Manderson informs us is the fact, should have for weeks, and sometimes for months, the highways leading to these central marts practically closed and an embargo placed upon the commerce of these cities and of all the region roundabout, because of the almost impassable condition of the public roads." * * *

ELOQUENT WORDS FOR GOOD ROADS.

The advantages to be derived from the system of permanently improved roads is so fittingly set forth by Prof. Pendergast, in an address before the Minnesota Good Roads Convention, that we quote his closing sentence: "Would not model roads be of greater benefit than our tobacco, liquor, tea, circus and other show moneys bring? Would they not more than balance the good times we have hauling our produce through miles of mud, at such fearful cost, in extra labor, repairs, horses, oats; in wear and tear of conscience and damages to character?

"It is certain that bad roads make weak, struggling churches and poor, ill-attended, lifeless schools. They necessitate a life of seclusion which walls the path of social progress.

"To sum up, a perfect highway is a thing of beauty and a joy forever. It blesses every home by which it passes. It brings into pleasant communion people who otherwise would have remained at a perpetual distance. It awakens emulation, cements friendships, and adds new charm to social life. It makes the region it traverses more attractive, the residences more delightful; it stimulates a spirit of general improvement. Fields begin to look tidier, shabby fences disappear, gardens show fewer weeds, lawns are better kept, the houses seem cosier, trees are planted along its borders, birds fill the air with music, the world seems brighter, the atmosphere purer. The country is awake, patriotism revives, philanthropy blossoms as selfishness fades and slinks from view. The school-house and the church feel the magic influence—the wand of progress has touched even them; the old are young again, the young see something new to live for, and to all life seems worth the living. The daily mail reaches each home. The rural cosmopolitan 'feels the daily pulse of the world.' Wheelmen are no longer confined to the cities. Bicycles, now within the reach of all, are no strangers among farmers. The golden days of which the poets long have sung are upon us. The dreams of the past are coming true. Nothing can thwart the will of fate. Put your ear to the ground even now and you will hear the footfalls of the 'good time coming.'"

PATERNALISM AND GOOD ROADS.

COL. RICHARD J. HINTON, IRRIGATION EXPERT, WASHINGTON.

"I am interested, sir, in the arguments made for, and the objections against, any aid or interference by the general Government. For myself I believe, not in paternalism, but in co-operation. We are still a government, for, by and through the people. I call the attention of those who oppose the resolution for an inquiry by the Department of Agriculture, on strict construction grounds, to the fact that the Constitution of the United States provides for the general Government's supervision and construction, too, of national highways, postal, military and wagon roads, and so on. Canals and railroad systems, too, have been the recipients of vast bounty. We have given for such building a land empire of not less than 200,000,000 acres of land. We have loaned the credit of this nation also to the extent of nearly or quite $100,000,000. I am not afraid, sir, to spell nation with a capital "N," even in the matter of road-making, because I am fortified by all the precedents. My personal experiences and observations run back to ante-bellum days, and I recall serving with road-surveying and road-making parties in the great West whose labors were paid for by the general Government. I am sure that I am well within the facts if I say that before the civil war began, and under strict construction administrations, this nation spent in the laying out and construction of highways, and in the surveying of routes for railroads, several million dollars. It was good money well spent. Our friends need not be alarmed. The paternalism that leads to the more common knitting together of the many communities that go to make up the American people is one to be encouraged. It is one of co-operation with, not command over, them.

"There is another point to which I will briefly refer. We are all made aware of widespread agricultural discontent. We hardly need be reminded that American history shows that such discontent has always been a portent, suggesting political changes and economic reconstructions. I am not going into these, but beg to call the attention of the National League for Good Roads to one important cause added to net production. What shall the farmer do with his surplus product, and why raise this surplus, if his way to the outside world is barred by impassable highways or obstructed by obstacles which increase threefold the expense of realizing the fruits of his industry? Can we enlighten him by showing him a better way of expending the taxes levied upon him for the improvement of his roads? Can we show him such a system, well driven home with clear persuasion and positive assurances, as will induce him to take hold of the subject with energy? I believe that the time is ripe to submit this matter to the intelligence of the American public. There should be a unity of interest between the city and the country. The city is almost as much interested in getting good roads as the country itself. To the country it means enhanced net value to country products; to the city it means a greater variety and no essential increase in cost to the consumer.

WILL FARM PROPERTY BEAR THE TAX NECESSARY TO BUILD GOOD ROADS?

John J. Whithall, of Woodbury, N. J., writes:

"The importance of the common road is perhaps realized by but few of us, and yet it is safe to say that of all the systems of intercommunication in our country of land and water there is none more important, none more wastefully expensive, and none more susceptible of improvement than the common roads. * * * 'There is no tax so great as the tax of bad roads,' and of all persons who have to pay such tax none have to pay it to so great an extent as does the farmer, for he has to haul all of his produce to market over these roads, and also nearly all of his supplies to his home over the same.

"A road may be described as a line of communication, and the ideal road is a line of the least resistance, level, straight and with a hard, smooth surface. The importance of this last will perhaps be more fully understood when we remember that the amount of power required to pull a load of one ton over different level surfaces requires greatly different powers, as the following table will show, together with the cost of moving the same:

"To pull one ton on sand requires 400 pounds, costing 40 cents; to pull one ton on hard earth requires 200 pounds, costing 20 cents; to pull one ton on macadamized road requires 50 pounds, costing 5 cents; to pull one ton on sheet asphalt requires 15 pounds, costing 2 cents; to pull one ton on iron tram-rails (street railway) requires 10 pounds, costing 1 cent; to pull one ton on steel rails (railway) requires 9 pounds, costing 9-10 cents; to pull one ton on water (canal) requires 2 pounds, costing 2-10 cents.

"If we examine this table the great difference in the amount of power required to pull a load over even the macadamized or stone road and over the iron tram or street railway will at once attract our attention, and would seem to point to the latter as the ideal road. * * *

"Since roads are *public* highways they, or at least the principal or leading ones, should be maintained by the *public;* that is, the main or principal lines of roads should be maintained by the county or State, the efforts of the local or township authorities being directed to the local roads. Under our State laws at present, upon application of two-thirds of the property owners abutting upon a certain road, offering to pay ten per cent. of the cost of a stone road over such highway, it becomes the duty of the Board of Freeholders of the county to have such road built, the cost, after deducting the ten per cent. which must be paid by the property owners abutting upon the road, to be paid one-third by the State and the other two-thirds by the county.

"Suppose, under this law, the county should, within the next twelve years, be called upon to build fifty miles of such road. We are assured that good macadamized roads, nine feet in width, have been built in adjoining counties at a cost of $4,000 a mile; but put the cost at $6,000, the cost of building the fifty miles of road would be $300,000; ten per cent. of this sum—$30,000—to be paid by the property owners abutting upon the road, leaving $270,000 to be paid by the county and State, of which the county would have to pay $180,000, which, distributed over the twelve years, would require $15,000 to be raised each year by taxation. The assessed value of the property in the county exceeds $14,800,000, so that, allowing for but a very slight increase in the taxables of the county, a tax of one dollar upon each $1,000 of the taxable property of the county for twelve years would be all the tax required to build us fifty miles of macadamized roads. * * *

"And would it not pay? Let each of us take the assessed value of our estates, and reflect if such a system of roads would not pay us more than one dollar per year for each thousand dollars for twelve years, required to build them. If it will not pay, how is it that England, France, Germany, Switzerland, and, in fact, all civilized countries, are building the best roads that can be made, and some of them in the face of difficulties that to us would be appalling. The Swiss, although far from being a wealthy people, have built roads through gorges and around precipices which would seem impossible; and which, it is stated, must have cost over $1,000,000 per mile. In England, where all the principal roads are improved, it is estimated that the saving effected by such improvement, so that three horses can do the work of four, amounts to $100,000,000 annually. In the State of Illinois it is stated that the cost of hauling the farm products of the State to market is at least $15,000,000 annually more than it would be if the roads were improved, and that such improvement would add $100,000,000 to the value of the farms. In one county in Northern Indiana, where the principal roads were macadamized by the county, the increased value of the farms, as valued by the farmers themselves, was nine dollars per acre, not for those farms upon the line of the improved road only, but the average increased value of the farms of the whole county. In our own State, wherever the improved roads have been built under this law, the price of farm lands has materially increased and the public are entirely satisfied with the expenditure, and continue increasing the number of miles built, considering it a good investment.

"Yet many farmers, when asked to favor a project for the improvement of roads, put it off, as they would a luxury, 'until better times,' while they acknowledge that better roads would be a good investment, just like raising higher grade stock, or using improved and labor-saving machinery. There are, we believe, very few farmers who, if they could, by procuring some labor-saving implements, effect near such a saving as would be made by improved roads, but would procure such implements at even a much greater cost than would be the cost to them of macadamizing the roads.

"There is still another aspect to the matter, the social one; improved roads would tend to destroy the isolation that is an objection to farm life, and promote sociability and intercourse among farmers and others; 'as iron sharpeneth iron, so the countenance of men sharpeneth that of his fellow-men,' the farmers would acquire new ideas, they would become more genial, and their laborious life become more pleasant. Man is a social animal, and as we mend our ways and give our city cousins the opportunity to come among us, and see more of us, as they pass along on our improved road they would rejoice at these evidences of our enterprise and perhaps conclude to cast their lot with us, and so build up our country and place it in the position its geographical position entitles it to. For these and for other reasons which might be adduced we believe we can answer the question in the affirmative, we believe that 'farm property can bear the tax necessary to build good roads.'"

ATLANTA EXPOSITION.

[*From the Public Ledger, Philadelphia, Pa.*]

THE O'DONNELL STEEL TRACK HIGHWAY RECEIVES FAVORABLE NOTICE AND THE RECOMMENDATION THAT SAMPLE SECTIONS BE BUILT.

Continuing the report of States as made at the Road Parliament lately held in Atlanta from the *Ledger* of Wednesday.

Besides the maps showing the progress made in road making in the different States, there was shown a collection of samples of the road materials found in each State. Another exhibit at the place of meeting was the model of a steel track highway exhibited by Hon. John O'Donnell, N. Y., and which attracted great attention both because of its construction and by reason of the generally expressed opinion that the steel track is the highway of the future. In the roadway, as the model is pictured, the track is made of rolled steel about one-quarter of an inch thick. It is not like a T rail, but a gutter track five inches wide on the inside, with a shoulder half an inch high and an outward and downward flange. This steel track is fastened to a longitudinal timber, three inches by six, which is spiked on cross ties. In fact, the road looks like an ordinary steel railroad, only the rail is a gutter track.

The fundamental condition for good roads is drainage, and the gutter track meets this, for the roadway between the tracks is of stone and rounded up in the center so the water runs off the road into the track, and wherever there is a depression in the road a side conduit takes the water to the side of the road. On this steel track a horse can draw twenty times as much as on a dirt road and five times as much as on a macadam, and it is the only road that offers a general system to extend all over the Union. The track will be uniform wherever laid, and if generally adopted can be traveled on from Portland to Mexico and from New York to Oregon. It seems to provide a perfect bicycle track and to be adapted to the horseless motor. The claim is that slow freight can be moved on this road cheaper than by the steam railroad, and that this road is to solve the problem of freight rates for both the farmer and the manufacturer. Between the tracks and at the sides the make-up is of dirt, stone, asphalt or other material.

On motion of Chief Consul I. B. Potter, of New York, the following resolution was unanimously passed:

Whereas, The exhibit and explanation of a steel track highway has aroused a substantial interest among the members of this Parliament; therefore

Resolved, That we recommend the construction of experimental lines of steel track highway at various points throughout the States for public travel in order that the practical value of such system may be determined.

IMPORTANT ACTION OF THE NEW YORK BOARD OF TRADE AND TRANSPORTATION.

GOOD ROADS AND HOW TO PROCURE THEM.

Adopted March 11, 1896.

The Second Annual Report of the Commissioner of Public Roads of the State of New Jersey says:

"The demand for improved roads is decidedly on the increase, especially in sections where they have been partially enjoying their benefits. Counties that have been slow in accepting the provisions of the State Aid Law are now anxious to avail themselves of it. Such is the demand throughout the United States for improved highways that politicians of all shades are making it a plea for popular favor, the platforms of both parties being pledged to liberal appropriations for improved roads, notably in Massachusetts, where the annual sum is approaching the million dollar mark."

Testimony from farmers is given in this report, of which the following are specimens:

Mr. Clayton Monroe, of Burlington County, N. J., says: "Before the construction of stone roads in Burlington County it cost fifteen cents per basket to market truck; now it costs three cents."

J. C. Whittall, of Woodbury, N. J., writes: "There is no tax so great as the tax of bad roads, and of all persons who have to pay such a tax none have to pay it to so great an extent as does the farmer, for he has to haul all his produce to market over these roads, and also a large part of his supplies for his home over the same."

The cost of pulling a ton a freight one mile is given in this report as follows: On sand, forty cents; on hard earth, twenty cents; on macadamized roads, five to ten cents, according to perfection of road; on iron tram rails, one cent; and in this connection the Commissioner speaks of wagon roads with steel wheel tracks as "the coming road."

Regarding this he states as follows:

"As all progress should be along the best lines, and as some of the best authorities predict, the coming highways will be of steel. It would be well for New Jersey, as she is the pioneer in State aid road improvement, to take the lead in inaugurating a system of steel roads, and thus ascertain by actual experience whether it is the most efficient and economical highway. The claims as presented are that the average cost of a macadam roadbed sixteen feet wide is about $7,000 per mile. The cost of a double steel track highway, sixteen feet wide, filled in between with broken stone, macadam size. is about $6,000 per mile. The cost of a rural one-track road, $2,000 per mile. The rails to be made of steel the thickness of ordinary boiler plate, and to be formed in the shape of a gutter. five inches wide with a square perpendicular shoulder half an inch high, then an angle of one inch outward, slightly raised. This forms a conduit for the water and makes it easy for the wheels to enter or leave the track. The advantages of steel rails are, first, longer wearing qualities than stone; second, one horse will draw on a steel track twenty times as much as on a dirt road, and five times as much as or macadam. We would therefore recommend that legislation be enacted and appropriations be made by which the Commissioner should be empowered to authorize the construction of experimental steel track roads."

The advantages of a steel track road, as stated in this report, are obvious, and would enormously increase the power embodied in farm animals; would extend the area from which our railroads receive produce for transportation to distant markets, enhance the value of real estate, benefit commerce and quicken the pulse of business wherever introduced.

A further obvious advantage of steel tracks is that they would greatly increase the durability of macadam or Telford roads by decreasing the wheel wear, which soon makes a depression in the wheel tracks, which the water on heavy grades rapidly deepens.

In cities we have a constant object lesson of the value of steel tracks by drivers of heavily loaded vehicles seeking the tracks of street railways to secure the easier traction, notwithstanding they are ill suited to the purpose, and often designed with high flanges to make it difficult for other vehicles to use them. If steel tracks are the coming road for rural districts, as suggested by the Road Commissioner of New Jersey, they are still better adapted for city pavements with heavy traffic, where the wheels soon wear grooves and holes in the pavements and crosswalks, necessitating frequent renewals. It is certain that steel tracks, suitable for ordinary vehicles, on the sides of streets now having street railroads, would remove the temptation for drivers to seek the railway tracks, enable street cars to make better time, and if laid down in other parallel streets prevent the concentration of freight traffic in passenger streets which now occurs—in other words, diffuse and equalize the movement of merchandise, instead of concentrating and congesting it in passenger streets as at present.

At the recent convention of American Wheelmen, in Baltimore, it was stated $1,200,000,000 were invested in farm horses and mules, and that for a considerable portion of the year 16,000,000 horses and mules stood in the stable, at an expense of $4,000,000 per day for horse feed, waiting for the mud to dry up, so that they might be employed. When the roads are bad local markets are illy supplied with produce, with the result of exorbitant prices to consumers, and when the roads dry up the farmers rush to town and glut the market in the first days of dry weather, with the result of abnormally low

Lightning Source UK Ltd.
Milton Keynes UK
UKHW012330061118
331891UK00010B/979/P